Radiation Imaging Detectors Using SOI Technology

Synthesis Lectures on Emerging Engineering Technologies

Editor
Kris Iniewski, *Redlen Technologies, Inc.*

Radiation Imaging Detectors Using SOI Technology
Yasuo Arai and Ikuo Kurachi
2017

Non-Volatile In-Memory Computing by Spintronics
Hao Yu, Leibin Ni, and Yuhao Wang
2016

Layout Techniques for MOSFETs
Salvador Pinillos Gimenez
2016

Resistive Random Access Memory (RRAM)
Shimeng Yu
2016

The Digital Revolution
Bob Merritt
2016

Compound Semiconductor Materials and Devices
Zhaojun Liu, Tongde Huang, Qiang Li, Xing Lu, and Xinbo Zou
2016

New Prospects of Integrating Low Substrate Temperatures with Scaling-Sustained Device
Architectural Innovation
Nabil Shovon Ashraf, Shawon Alam, and Mohaiminul Alam
2016

Advances in Reflectometric Sensing for Industrial Applications
Andrea Cataldo, Egidio De Benedetto, and Giuseppe Cannazza
2016

Sustaining Moore's Law: Uncertainty Leading to a Certainty of IoT Revolution
Apek Mulay
2015

Radiation Imaging Detectors Using SOI Technology

Yasuo Arai and Ikuo Kurachi

ISBN: 978-3-031-00905-1   paperback
ISBN: 978-3-031-02033-9   ebook

DOI 10.1007/978-3-031-02033-9

A Publication in the Springer series
SYNTHESIS LECTURES ON EMERGING ENGINEERING TECHNOLOGIES

Lecture #9
Series Editor: Kris Iniewski, *Redlen Technologies, Inc.*
Series ISSN
Print 2381-1412   Electronic 2381-1439

Radiation Imaging Detectors Using SOI Technology

Yasuo Arai and Ikuo Kurachi
High Energy Accelerator Research Organization (KEK)

SYNTHESIS LECTURES ON EMERGING ENGINEERING TECHNOLOGIES #9

ABSTRACT

Silicon-on-Insulator (SOI) technology is widely used in high-performance and low-power semi-conductor devices. The SOI wafers have two layers of active silicon (Si), and normally the bottom Si layer is a mere physical structure. The idea of making intelligent pixel detectors by using the bottom Si layer as sensors for X-ray, infrared light, high-energy particles, neutrons, etc. emerged from very early days of the SOI technology. However, there have been several difficult issues with fabricating such detectors and they have not become very popular until recently.

This book offers a comprehensive overview of the basic concepts and research issues of SOI radiation image detectors. It introduces basic issues to implement the SOI detector and presents how to solve these issues. It also reveals fundamental techniques, improvement of radiation tolerance, applications, and examples of the detectors.

Since the SOI detector has both a thick sensing region and CMOS transistors in a monolithic die, many ideas have emerged to utilize this technology. This book is a good introduction for people who want to develop or use SOI detectors.

KEYWORDS

radiation image sensor, silicon-on-insulator, SOI, X-ray diffraction, X-ray imaging, X-ray astronomy, high-energy particle physics, synchrotron radiation, pixel detector, CMOS, radiation tolerance, monolithic sensor, particle detector

Contents

Preface

Currently there are many kinds of smart silicon sensors around us such as optical cameras, temperature/pressure sensors, accelerometers, and so on. However, most X-ray sensors are still a composition of a bulky scintillator and optical device. Since the X-ray energy is detected through scintillation light (indirect detection), position and energy resolutions are limited. To get better performance, semiconductor sensors which directly detect electrons or holes generated by radiation have been developed. In most cases, both the sensors and readout electronics use silicon as a material, but they are produced in different substrates and bump bonded as a hybrid detector.

In modern high-energy physics experiments, silicon pixel detector is one of the most important detectors. It measures the track position of secondary particles which are created by the high-energy particle collisions, in micron order precision. Development of the pixel detector was started in the 1990's and widely used in the Large Hadron Collider (LHC) experiments at CERN. However, these detectors were also hybrids of Si sensor and LSI readout chip. Therefore, they are very expensive and hard to fabricate due to its millions of bumps.

Many people have tried to develop a monolithic radiation detector which has both sensors and electronics in a same die. Silicon-on-Insulator (SOI) technology was the first candidate to realize these demands. However, lack of technology and coupling effect between sensors and electronics make it difficult to realize.

The Japanese high-energy accelerator laboratory, KEK, was behind in the development of pixel detectors as were many semiconductor companies in Japan. Several Japanese companies has been involved in the development of SOI wafer and LSI process from the beginning. NTT LSI lab. developed the first commercial-level SOI wafers by using SIMOX process and OKI Electric Industry Co. Ltd. (later Lapis Semiconductor Co. Ltd.) was the first worldwide to mass produce SOI devices.

To catch up to the radiation pixel detector technology developed in Europe, we started an SOI pixel detector project in 2005 in cooperation with the Lapis. At that time, there were very few researchers with experience in silicon detector or design of LSI circuits, and people in the Lapis didn't have any experience in radiation detection. Furthermore, the market of radiation image detector is not so large, so there were several difficult times to continue the cooperation. However, enthusiasm of the involved people enabled us to complete the SOI radiation detector.

Since the initial cost of the semiconductor process is very expensive, we decided to open our process to all academic users worldwide. We have been operating Multi Project Wafer (MPW) runs, by collecting many designs, one or twice per year. In addition to Japanese researchers, several foreign researchers also joined the MPW runs. Coordination of the MPW run is not an easy task, but we believe it is important task to continue the SOI detector project.

Recognized for our activity, the Japanese government gave us some R&D funds for the 2013–2017 fiscal year. Our collaboration was expanded with this R&D budget, and now we can bring electronics and image sensor experts to our collaboration. Furthermore, the collaboration could involve researchers in other fields such as mass separator, medical, material science, and so on.

The First International Workshop on SOI Detector (SOIPIX2015) was held in June 2015, when about 100 people from all over the world gathered in Sendai, Japan. We hope this book will help to expand SOI pixel technology worldwide.

SOIPIX2015 workshop in Sendai, Japan (June 3-5, 2015).

Yasuo Arai and Ikuo Kurachi
January 2017

Acknowledgments

The development of the SOI pixel detector was started as a project of the Detector Technology Project which was initiated by Prof. Takasaki and Prof. Haba of High Energy Accelerator Research Organization (KEK). We acknowledge their continuous support to this SOI detector project.

We gratefully acknowledge people at Lapis Semiconductor Co., Ltd, especially Mr. Okihara, Mr. Kasai, Mr. Miura, and Mr. Kuriyama. Without their continuous efforts, the SOI detector would not have been built. We also thank all the members of SOI collaboration who were involved in many aspects of work, such as simulation, design, testing, and discussion.

This work was supported by MEXT KAKENHI Grant Number 25109001~9. This work is also supported by LSI Design and Education Center (VDEC), the University of Tokyo with the collaboration with Cadence Corporation, Synopsys Corporation, and Mentor Graphics Corporation.

Yasuo Arai and Ikuo Kurachi
January 2017

CHAPTER 1

Introduction

Silicon pixel detectors are indispensable tools for modern reasearch in high-energy physics [1], medical equipments [2], space science [3], and many other fields. At present, the most popular radiation image sensor is built as a hybrid of sensor chip and readout IC chip with metal bump bonding [4, 5]. However, the hybrid sensor has many limitations in performance, such as position resolution, thick unwanted materials, low production yield, and high cost. Therefore, a monolithic active pixel detector that has both sensors and a readout circuit in a single silicon die and is fabricated through a conventional semiconductor process is demanded.

There are several kinds of monolithic sensors. Using a standard bulk Complementary Metal-Oxide-Semiconductor (CMOS) process is the most convenient and widely used method. However, it is difficult to make thick sensing regions in the standard CMOS process [6], and it does not necessarily have the desired properties such as radiation hardness. Silicon-On-Insulator (SOI) technology becomes popular for high-performance, RF, or low-power IC applications. In SOI wafer, the circuit layer is isolated from the substrate (handle wafer). The substrate is normally just a physical support of the wafer and just acts as a backplane of the top circuit. However, the handle wafer is also a good quality silicon wafer, so active elements or sensors can be created in the handle wafer. To get good sensitivity for X-ray or infrared light, a thick sensing region and low-leakage current is mandatory. This necessitates using a high-purity silicon wafer different from low-resistivity wafer used in the circuit layer. To detect α-ray, β-ray, or high-energy charged particles, back illumination and thin window are normally required. These requirements are also difficult to fulfill in standard bulk CMOS technology.

The idea of the SOI detector first appeared around the 1990s [7]. At first, the SOI wafer was fabricated with a separation by implantation of oxygen (SIMOX) technology [8] from low-resistivity p-type wafer. Therefore, the detector could not provide advantages of the fully depleted one. Several pioneering works were done [9–11] but it was difficult to make a good detector due to insufficient technology in the SOI wafer and process.

In 2001, the development of SOI detectors with high-resistivity handle wafer was started by a collaboration of Silicon Ultra Fast Camera for Gamma and Beta Sources in Medical Applications (SUCIMA) project [12–14]. It was the first successful work on SOI detectors with high-resistivity handle wafer. The handle wafers were manufactured by using bond-and-etch-back SOI (BESOI) technology. Unfortunately, the process technology used was rather obsolete (CMOS 3 μm technology), and it suffered from many technical problems.

A Japanese group led by KEK started SOI Pixel R&D in collaboration with Lapis Semi-conductor Co., Ltd. (former company name was OKI Electric Industry Co. Ltd.) [15] in 2005. The company was the first supplier of the mass-produced SOI Large-Scale Integrated circuit (LSI) in the world [16, 17]. They have been using bonded wafers called UNIBOND™ wafers made by Smart Cut™ method [18–20], in which high-resistive and low-resistive wafers can be bonded, sandwiching a buried oxide (BOX) layer. The thin low-resistivity upper Si layer is used for circuit implementation, and the high-resistivity bottom wafer (handle wafer, substrate) is used as a sensor. At first the pixel process is developed using R&D line of 0.15 μm CMOS Fully-Depleted (FD)-SOI process [15], and then moved to 0.20 μm FD-SOI process of mass-production line [21, 22]. The group has been operating Multi Project Wafer (MPW) runs, open to academic users worldwide, approximately twice per year [23–25]. They have solved many issues for realizing SOI radiation detector in SOI technology and developed many kinds of detectors. Furthermore, to fulfill demanding requirements in actual experiments, many new techniques, such as nested well, double-SOI, stitching, and 3D vertical integration [26], are being developed. A review article exists about the past activities on SOI development [27].

Figure 1.1 shows the schematic view of the SOI pixel detector (SOIPIX). The SOI wafer is composed of a thick, high-resistivity substrate (sensor part) and a thin low-resistivity Si layer (CMOS circuitry) sandwiching a buried oxide (BOX) layer. After removing the top Si and the BOX layer in the region of the sensing node contacts, p or n dopant is implanted to the substrate. Then contact vias and metal connections from the p-n junction to the transistors are created. The main advantages of the SOI detectors are as follows.

- There is no mechanical bump bonding, so obstacles, which will cause multiple scattering, are eliminated and smaller pixel size is possible.

- Parasitic capacitances of sensing nodes are very small (~10 fF), so large conversion gain and low noise operation are possible.

- Full CMOS circuitry can be implemented in the pixel.

- The thickness of the sensing region can be adjusted for radiation species and application requirement.

- Since generated electron-hole drift with electric field, position resolution is very good as high as a few micron. This is far better than X-ray sensor using scintillator where scintillation light is emitted at 360°.

- The cross section of single event effects caused by radiation is very small. A latch-up mechanism, which destroys conventional bulk CMOS LSI, is absent.

- Unlike conventional CMOS process, there is no leakage path to bulk. Thus, SOI transistors are shown to work over a very large temperature range from below 1~600 K.

- The technology is based on industry standards, and one of most promising technology for future LSIs. Thus, further progress and lower cost are foreseeable.

- Emerging vertical (3D) integration techniques are a natural extension of the SOI technology, so a much higher integration density is possible.

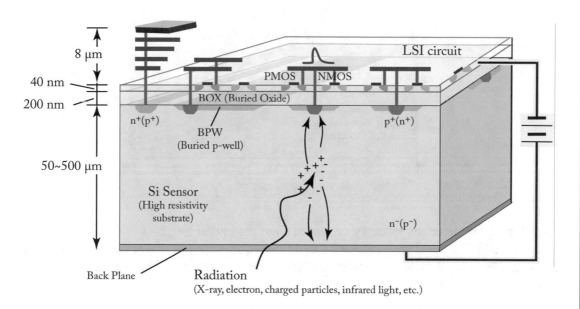

Figure 1.1: Schematic view of the typical SOIPIX.

The type of handle wafer can be either *n* or *p*, and doping impurity is changed according to the substrate type. We assume *n*-type substrate here otherwise noted.

To have good sensitivity, the depletion layer must have enough thickness and be depleted fully. Figure 1.2 shows X-ray detection efficiency for different silicon thickness. By using 500 μm thick silicon in full depletion condition, almost 100% detection efficiency is available for 10 keV X-ray, and 15% for 30 keV. When an X-ray interacts with silicon, a number of electron-hole pairs of (X-ray energy)/(3.65 eV) are created on average, so about 2700 e-h is available for 10 keV X-ray. For 1 keV X-ray, only 270 e-h is created and penetration depth is a few micro meters, therefore it is important to have a thin entrance window and low electronics noise. On the other hand, about 80 e-h/μm is created along with the track of high-energy charged particle. Thus, about 4000 e-h is available with only 50 μm thick sensor.

To deplete thick sensor with lower voltage, the dopant concentration in the handle wafer must be as low as 10^{12} cm^{-2}. For example, 200 V is necessary to fully deplete 10^{12} cm^{-3} (~4.5 kΩ•cm) *n*-type wafer of 500 μm thick. On the other hand, the dopant concentration in SOI circuit should be kept enough high as 10^{15} cm^{-3} to maintain the MOSFET characteristics. Then, the SOI wafer made by wafer bonding method with Smart Cut™ is preferable in the SOIPIX.

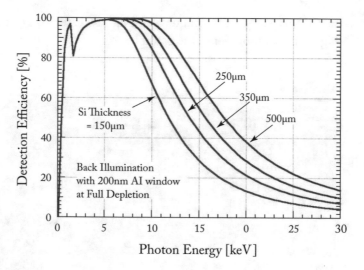

Figure 1.2: X-ray detection efficiency calculated from absorption length of *Si* sensor for different silicon thickness. The calculation assumes backside illumination through 200 nm Aluminum window and no dead layer in *Si*. For actual detection efficiency in a low-energy region, signal-to-noise ratio also limits the efficiency.

In Chapter 2, issues inherent in SOI pixel implementations are summarized, then in Chapter 3 technology to solve these issues mainly developed by KEK and LAPIS Co. Ltd. are introduced. In Chapter 4, we discuss radiation hardness of the SOI devices. In Chapter 5, recent R&D to expand the possibility of the SOIPIX such as stitching exposure and 3D integration are shown. In Chapter 6, various detectors developed so far are described.

CHAPTER 2

Major Issues in SOI Pixel Detector

In this chapter, major issues to realize SOI radiation detectors are presented. These are back-gate effect, crosstalk, leakage current, high-resistivity wafer process, and radiation hardness. Process modifications to solve these issues are described in Chapter 3 and radiation hardness issues are discussed in Chapter 4.

2.1 BACK-GATE EFFECT

The back-gate effect is one of the major difficulties with building a radiation sensor in SOI wafer [28–30]. Transistors are located very closely (~200 nm) to the sensor where high voltage is applied. The potential under the BOX acts as a back gate of the transistors. As the back-gate voltage is increased, the threshold voltage of the n-channel transistor (NMOS) is decreased and that of the p-channel transistor (PMOS) is increased, which will in the end cause the circuit not to work. Figure 2.1 shows an example of the threshold shifts in 0.15 μm SOI process.

2.2 CROSSTALK BETWEEN SENSORS AND CIRCUITS

For the SOI pixel detector, the BOX layer separates the electronics circuit and the sensor. Since the thickness of the BOX is very thin (~200 nm), voltage change in circuitry can induce signals in the sensors through capacitive couplings, and then signal oscillation may occur in some cases. Coupling path depends on the layout of pixel. Unfortunately, buried p-well (BPW) (Section 3.2.3) introduced to reduce the back-gate effect increases the capacitive coupling.

In integration-type pixels shown later, this coupling does not have any serious effect on the measurement. However, in counting-type pixel, this may have a large effect. In this case, differential signals and/or some sort of shielding between the sensors and circuits must be introduced. Thus, the nested well structure (Section 3.2.3) and double SOI technique (Section 5.1) are proposed and developed.

2.3 LEAKAGE CURRENT

During measurement, a leakage (dark) current flows into the sense node from the diode fabricated in the handle wafer. This increases the output voltage and noise. The leakage current of the sensor

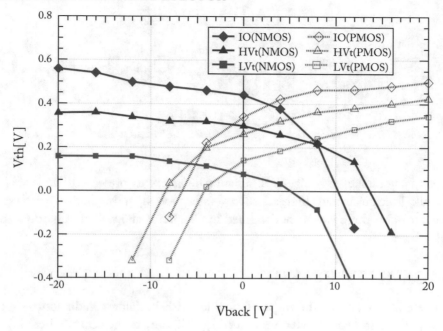

Figure 2.1: Measured threshold Voltage (Vth) shifts due to the back-gate voltage (Vback) for NMOS and PMOS transistors of the three different transistor type in 0.15 μm SOI technology.

is not so important for short integration time applications, but it is important for long integration time applications such as astronomical X-ray observation and measurements that require very good energy resolution.

The major leakage current comes from generation current in the depletion region, bonding interface between the BOX and substrate, and carrier diffusion from un-depleted region. These values strongly depend on manufacturer and wafer type. Low-temperature operation and buried p-n diode will help a lot to decrease the leakage current.

2.4   HIGH-RESISTIVITY WAFER

To have high-sensitivity for radiations and to get high-speed signal, a sensor must be depleted fully and excess voltage applied to speed up charge collection. The depletion depth is proportional to square root of $\rho \cdot V$ where ρ is resistivity of the wafer and V is applied voltage. Therefore, high-resistivity wafer is desirable to achieve full depletion with lower voltage. Usually Floating-Zone (FZ) wafer is used in a radiation sensor, but it is not easy to obtain large-size FZ wafers because of their processing difficulty and demand.

The handle wafer of the standard high-resistive SOI wafer is made in the Czochralski (CZ) method, which is n-type and has about 700 $\Omega \cdot$cm resistivity. KEK group made special SOI wafers by bonding FZ wafer for both *n*- and *p*-types to meet the needs of SOI detector research. To do CMOS process on FZ -SOI wafer is difficult task since CMOS high temperature process will cause slips in the wafer, but this is solved by modifying the high temperature process, as described in Section 3.2.1

2.5 RADIATION HARDNESS

Radiation hardness is one of the crucial performance issues for particle and X-ray detectors. The SOI device is known as a radiation hard device for Single Event Effect (SEE) since the active *Si* thickness of the SOI transistor is very thin and the number of generated charges in the active area is very small compared with bulk transistors. Thus, higher immunity to SEE is obtained in a SOI device, and many SOI devices have been used in satellite instruments [31].

But, due to the relatively thick silicon dioxide layer, BOX, the SOI device is more sensitive than that of bulk silicon device in Total Ionization Dose (TID). As BOX is exposed to ionizing radiation, radiation-induced charge will be trapped in the BOX. This radiation-induced trapped charge is predominantly positive. And this charge buildup in the BOX can invert the back-channel interface, forming a leakage path between the source and drain of the top-gate transistor. This leakage current resulting from the radiation-induced charge buildup in the BOX will prevent the top-gate transistor from being completely turned off. There are many experiments [32, 33] that assess the total dose effect from ionizing radiation performed on FD-SOI detectors, and they showed how the substrate bias condition during irradiation plays a dramatic role on the resulting radiation damage.

CHAPTER 3

Basic SOI Pixel Process

3.1 ADVANTAGES OF SOI STRUCTURE

A bulk CMOS structure is commonly used in most semiconductor devices such as logic and memory large-scale integrated circuits (LSI) [34]. The schematic cross section of bulk CMOS structure is shown in Fig. 3.1a. The bulk CMOS structure consists of n-type and p-type MOSFETs. The p-well for n-MOSFETs and n-well for p-MOSFETs are formed in the p-type silicon substrate. The bulk CMOS structure has three disadvantages especially for future LSIs. The first disadvantage is "Latch-up" problem because of parasitic PNPN thyristor formation in silicon substrate. The P or N well current generated by drain avalanche during the MOSFET operation, injected current from outside, or radiation induced current can trigger the thyristor and consequently "latch-up" or logic miss-operation occurs. The second one is relatively high parasitic capacitance of source and drain because of high permittivity of silicon in depletion region. The parasitic capacitance causes slow operation or high power consumption. The last one is higher threshold voltage variation because of high channel doping for advanced LSIs to reduce short channel effect. To overcome these disadvantages, SOI CMOS structure is the most suitable.

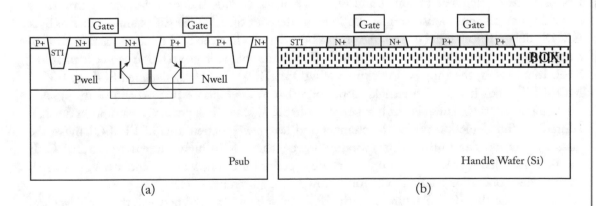

Figure 3.1: (a) Schematic cross section of conventional bulk CMOS structure. (b) Schematic cross section of SOI structure.

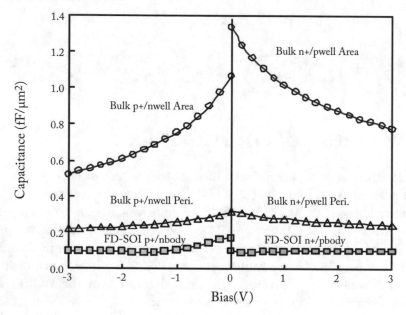

Figure 3.2: Junction (source and drain) parasitic capacitance for bulk CMOS and SOI [38].

The SOI CMOS structure is shown in Fig. 3.1b. The oxide layer called buried oxide (BOX) separates between active MOSFET (SOI) layer and a handle wafer. Individual MOSFET is perfectly isolated by surrounding oxide. Then, no latch-up issue exists because of no thyristor and low parasitic capacitance of source and drain is realized (Fig. 3.2) due to the lower permittivity of oxide than silicon. Consequently, the inverter delay of SOI MOSFET is almost half of bulk CMOS delay. The channel concentration can be reduced when the SOI thickness is thin enough since the gate can easily control the channel. Then, threshold voltage variation can be kept at low level. In addition, the thin SOI thickness of MOSFET, which is fully depleted SOI (FD-SOI) MOSFET, leads lower subthreshold slope than that of the bulk CMOS, as shown in Fig. 3.3. This characteristic is suitable for low supply voltage, V_{cc}, and low power consumption devices. Moreover, full depletion under the channel and less *p-n* junction area of FD-SOI make the device operation from ultralow temperature lower than 1 K to high temperature as 250°C. It can be concluded that the FD-SOI devices are superior for ultralow power and low V_{cc}, or harsh environment operation such as radiation or low and high temperatures.

To make the SOI structure, only the BOX must be introduced between the SOI layer and the handle wafer, as shown in Fig. 3.1b. Another requirement is tight thickness control of the SOI layer. There are two common techniques to form the SOI structure. One is Separation by IMplantation of OXygen (SIMOX) [35] and the other is bonded SOI using the Smart Cut[TM] technology [36, 37]. SIMOX technology uses high-dose oxygen implantation underneath active

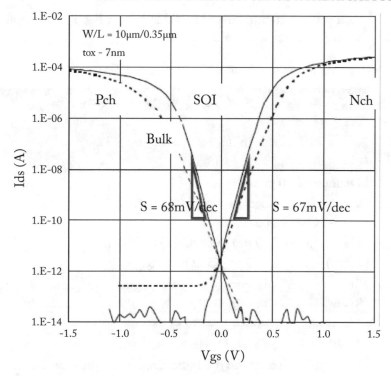

Figure 3.3: Subthreshold characteristics for bulk CMS and SOI [38].

silicon layer. After the implantation, high-temperature annealing at around 1300°C is employed to recrystallize upper silicon layer and to form stoichiometric oxide as BOX.

For bonding technique with the Smart Cut[TM] [20], two wafers are needed to be prepared. One is for the handle wafer and the other is for the SOI active layers. The wafer of SOI layer is oxidized and implanted hydrogen with dose range of 10^{16} cm^{-2}. The energy should be selected that the projection range of implanted hydrogen is the same as the target thickness of SOI active layer. Then these two wafers are bonded by Van der Waals force. The cleave plane at the hydrogen implanted region is easily formed after around 500°C annealing and the top region silicon wafer is cut away. Finally, high temperature annealing at around 1100°C is performed to strengthen bonding of two wafers. One of merits using the SOI wafer by bonding with the Smart Cut[TM], is that two different wafers can be used to form the SOI structure.

3.2 FD-SOI RADIATION SENSOR FABRICATION PROCESS

The process is an important factor to limit the development of SOI detector. SOI pixel process was developed based on 0.15 μm CMOS FD-SOI process [15] and later moved to 0.2 μm process [21] in collaboration with LAPIS Semiconductor Co. Ltd. Main specifications of the

KEK SOI CMOS pixel process is summarized in Table 3.1. In Fig. 3.4, a simplified procedure for the fabrication of the SOI pixel detector is shown.

Table 3.1: KEK SOI pixel process specifications

Objectives	Details
Process	0.2 μm Low-Leakage Fully-Depleted SOI CMOS, 1 Poly, 5 Metal layers, MIM capacitor (1.5 fF/μm2), DMOS option. Core (I/O) Voltage = 1.8 (3.3) V
SOI Wafer	Diameter: 200 mmφ, Top Si: Cz, ~18Ωcm, p-type, ~40 nm thick Buried Oxide: 200 nm thick Handle wafer: 720 μm thick. Cz(n) ~700Ωcm, FZ(n) ~7kΩcm, FZ(p) ~25 kΩcm, etc.
Backside	Thinned to 50~500 μm by mechanical grind, and chemical etching. Then adequate impurity is implanted and laser Annealed. Then Al is deposited (200 nm).
Transistors	Multi Vt for core and IO transistors. Four types of structures (body-floating, source-tie type 1~2 and body-tie) are available.
Optional Process	Several kinds of buried n/p-well formation, Vertical integration with μ-bumps.

The base SOI process has been developed for ultralow power operation application by LAPIS semiconductor Co., Ltd. The process employs 1 poly and 5 metal layers with LOCOS isolation, dual gate oxide for 1.8 and 3.3 V operation, poly and diffusion Co-salicide for reduction of parasitic resistance and making contacts reliable [39, 40], W filled via, and MIM capacitors. The thickness of BOX and SOI are 200 nm and 40 nm, respectively.

Additional process steps for sensor fabrication are added to the base process. For X-ray sensor fabrication, $p+/n-$ junction diode formation in the handle wafer is newly added to the commercial FD-SOI process [41]. The major processes dedicated for the X-ray sensor are: (i) thermal process optimization for usage of ultralow doping concentration floating zone (FZ) wafer as the handle wafer; (ii) $p+$ (or $n+$) diffusion layer formation on the surface of the handle wafer and contact formation; and (iii) well formation in the handle wafer. These three items will be explained later.

As shown in Fig. 3.4, the backside electrode must be formed to bias high positive voltage to the $p+/n-$ junction diode. After the front side process, the backside grinding is performed to get a given wafer thickness. To remove the damaged layer by grinding, chemical wet etching

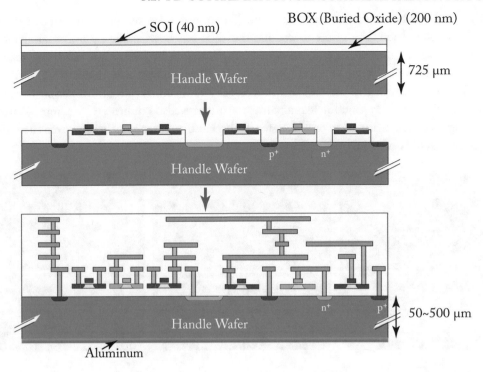

SOI (40 nm) BOX (Buried Oxide) (200 nm)

Handle Wafer 725 μm

Handle Wafer p⁺ n⁺

Handle Wafer n⁺ p⁺ 50~500 μm

Aluminum

Figure 3.4: Simplified SOI pixel process flow.

is employed. For reduction of contact resistance, phosphorus ions for *n*-type substrate (boron ions for *p*-type substrate) are implanted from backside. Activation of the implanted dopants is performed by laser annealing. Then, typically 200 nm aluminum film is deposited by vacuum evaporation to make a backside electrode.

3.2.1 FZ WAFER UTILIZATION FOR HANDLE WAFER

As described above, the *p-n* junction of X-ray sensor is consisted in the handle wafer. Therefore, requirements of the handle wafer are ultralow doping concentration with less concentration variation and low crystal defects. CZ grown silicon wafers are commonly used in the LSI process. However, the CZ wafer contains relatively high concentration of oxygen due to its manufacturing method. The oxygen works as a thermal donor in silicon and low doping concentration control is difficult. It is also well known that the oxygen precipitation during high-temperature annealing leads bulk micro defects which cause junction leakage. To reduce oxygen concentration in silicon, FZ grown silicon wafer should be the best material as handle wafer in SOI structure. However, mechanical strength of FZ wafer is low and crystal defects such as slips can be generated after high-temperature annealing.

The SOI-CMOS fabrication process itself contains high-temperature annealing, as described previously. Then, the effort to reduce the slips has been made by optimizing recipe of such high-temperature annealing. Results of X-ray topography before and after the recipe optimization are shown in Fig. 3.5. The slips after the high-temperature annealing (vertical and horizontal white lines in the picture) are successfully reduced by the optimization of ramp-up and down conditions. The junction leakage improvement is also confirmed by using FZ wafer, as shown in Fig. 3.6. The utilization of FZ wafer as the handle wafer greatly improves the radiation sensor characteristics of $p+/n-$ junction diode.

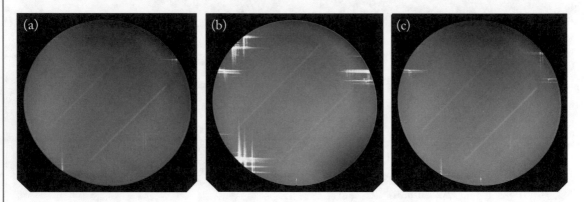

Figure 3.5: X-ray topographies of high temperature annealed SOI wafer with FZ handle wafer: (a) before oxidation, (b) conventional process, and (c) after process improvement.

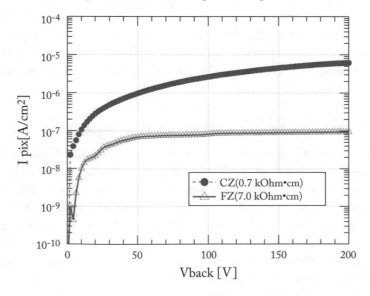

Figure 3.6: Junction leakage current of $p+/n-$ junction diode in a pixel for CZ and FZ handle wafer.

3.2.2   *p-n* JUNCTION DIODE AND CONTACT

It is not difficult to make $p+/n-$ junction diode in the handle wafer. After FD-SOI MOSFET fabrication and before the wiring process, oxide etching at the area where the $p+$ diffusion is made is performed by conventional lithography and etching process. Boron ions are implanted into this hole prior to the thermal activation annealing. The contacts to the diffusion layer are opened by using the same process conditions for the conventional SOI devices. All contacts including the contacts to $p+/n-$ junction diode diffusion layer and SOI device contacts are filled by tungsten CVD. Then, fine structure to contact the $p+/n-$ junction diode is feasible.

3.2.3   WELL FORMATION IN HANDLE WAFER

As mentioned above, the high positive voltage bias is applied to the handle wafer to make wider depletion region. This bias alters the MOSFET characteristics because of the floating body FD-SOI MOSFET known as a back-gate effect. Id-Vg characteristics of n-MOSFET with the substrate bias is shown in Fig. 3.7a. As shown in the figure, the threshold voltage decreases with the bias and the off-current reaches around 1×10^{-3} A when the bias is 50 V. The bias must be over 100 V to make fully depleted $p+/n-$ junction diode and the circuits in SOI don't work correctly in such a high bias.

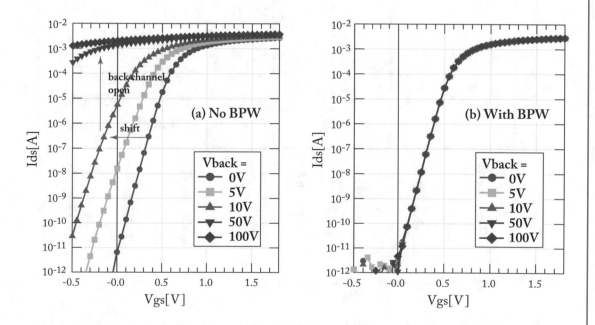

Figure 3.7: NMOS I/O transistor Ids-Vds curve ($L = 0.35\ \mu$m, $W = 175\ \mu$m, @$Vds = 0.1$ V) and back side voltage. (a) Without the BPW layer, and (b) with the BPW layer connected to ground.

To suppress this back-gate effect mentioned in Chapter 2, *p*-well in the handle wafer which is known as buried *p*-well (BPW) is newly introduced underneath the SOI circuits. When the BPW is biased to the ground, the back-gate effect is totally eliminated, as shown in Fig. 3.7b. Then, the basic operation of FD-SOI X-ray sensor is confirmed with this BPW process [42, 43].

The introduction of the BPW is realized by boron ion implantation through the SOI layer and BOX. The boron ions pass through the SOI layer because SOI thickness is only around 40 nm and BOX thickness is 200 nm. The effects of the BPW formation to MOSFET characteristics are also investigated by comparing threshold voltages of MOSFETs with or without the BPW implantation, as shown in Fig. 3.8. Threshold voltage difference is almost within 20 mV and the effect to MOSFETs can be negligible.

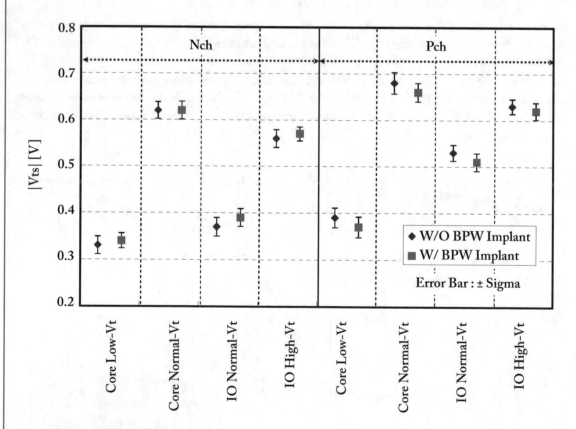

Figure 3.8: Threshold voltage comparison between with and without the BPW.

It can be used to shield transistors outside the pixel such as at periphery. It will also shield transistors in the pixel area since it normally connected low voltage node, but it increases the coupling between the sensing node and circuits. The BPW is not providing direct benefit for protection against charging of the BOX layer as a result of accumulation ionizing doses of the incident radiation. But, a less electric field in the BOX improves radiation hardness.

This process can be extended to the formation of more complicated well structure such as well-in-well in the handle wafer. The example of the BPW and well-in-well is shown in Fig. 3.9 [44]. All of the wells are formed by ion implantation through SOI and BOX layers.

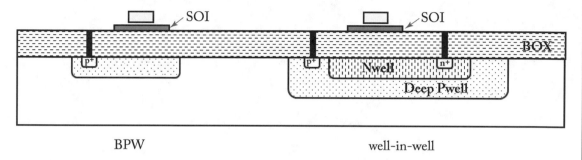

Figure 3.9: Various buried well structures in this technology.

The BNW layer is connected to a fixed voltage to shield transistors from the bottom. The well-in-well, so-called nested well, structure also reduces crosstalk between the sensors and circuits but it needs additional area to make the structure and the capacitance between the nested layers a bit high (~2 pF for 105 μm × 105 μm area at 1 V).

CHAPTER 4

Radiation Hardness Improvements

4.1 FD-SOI RADIATION HARDNESS

It is well known that FD-SOI MOSFET has much better tolerance to radiation-induced single event effect (SEE) than bulk CMOS. However, total ionizing dose (TID) degradation is one of crucial issues for the FD-SOI [45]. In this section, the causes of X-ray irradiation damage of the FD-SOI n- and p-MOSFET are investigated in detail. Based on the investigation results, improvement methods for the radiation hardness are also proposed.

Figure 4.1 shows the linear region drain current change by the X-ray irradiation for n-MOSFET and p-MOSFET. After 1.4 kGy(Si) irradiation, the change is around 15% increase for n-MOSFET and 20% decrease for p-MOSFET. Improvement of the radiation hardness has to be achieved at least more than 10–20 kGy(Si) within 10% drain current change based on radiation experiment requirements. To improve the radiation hardness, the causes of drain current change by the X-ray irradiation must be analyzed. In general, it is reported that the X-ray irradiation degradation is caused by the positive charge generation in oxide and the interface state generation between oxide and silicon [46].

In the FD-SOI case, there are two oxide films, which are BOX and gate oxide and two interfaces between gate oxide and SOI and between BOX and SOI. These two oxide films and two interfaces should be paid attention. In the case of n-MOSFET, the drain current increases with the X-ray irradiation. Then, this may be caused by the positive charge in gate oxide or BOX. To improve the radiation hardness of n-MOSFET, we have to distinguish which is the dominant cause of drain current change, charge in gate oxide or charge in BOX. In the case of p-MOSFET, the drain current decreases. Then, the suspected causes are: (i) zero substrate bias threshold voltage $|V_{to}|$ increase by the charge in gate oxide; (ii) positive back-bias by the charge in BOX; (iii) local $|V_{to}|$ increase by the charge in sidewall spacer; and (iv) mobility reduction by the generated interface states. Because there are four suspected causes in p-MOSFET drain current change, new analysis method is required to identify the cause.

4.2 CAUSE OF n-MOSFET DRAIN CURRENT CHANGE

The possible causes to increase n-MOSFET drain current are the positive charge in gate oxide or BOX. To distinguish the causes, the analysis using maximum trans-conductance dependence

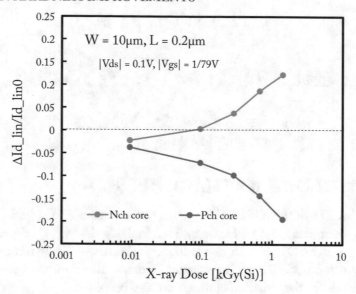

Figure 4.1: Drain current change for n- and p-MOSFET as a function of X-ray dose.

on substrate bias ($g_{mmax} - V_{sub}$) is proposed. The typical $g_{mmax} - V_{sub}$ characteristic is shown as a solid line in Fig. 4.2a. After the X-ray irradiation, the $g_{mmax} - V_{sub}$ curve may shift as dashed line in the figure. Then, the number of trapped holes in BOX, N_{ot_box}, is calculated from ΔV_{sub} as

$$N_{ot\_box} = \frac{\epsilon}{q T_{ox\_box}} \Delta V_{sub}, \tag{4.1}$$

where T_{ox_box} is the BOX oxide thickness, q is the elementary electron charge, and ϵ is the permittivity of oxide. The number of trapped holes in the gate oxide, N_{ot_gox}, is also calculated from

$$N_{ot\_gox} = \frac{\epsilon}{q T_{ox\_gox}} \left[V_{tt\_deg} (0 + \Delta V_{sub}) - V_{tt\_int} (0) \right], \tag{4.2}$$

where $V_{tt\_int}(x)$ is the threshold voltage at $V_{sub} = x$ for the fresh device and $V_{tt\_deg}(x)$ is that for the irradiated device.

The $g_{mmax} - V_{sub}$ characteristics for each X-ray dose are shown in Fig. 4.2b. The calculated number of trapped holes in gate oxide and BOX is shown in Fig. 4.3a as a function of X-ray dose. Based on the results of Fig. 4.3a, calculated ΔV_{to} is shown in Fig. 4.3b. As shown in the figure, V_{to} shift by the generated charge in gate oxide is only 30 mV but that by the generated charge in BOX is 130 mV. Therefore, the generated charge in BOX is the major cause of the drain current change in n-MOSFET. To improve the radiation hardness of n-MOSFET, the charge generation in BOX must be reduced.

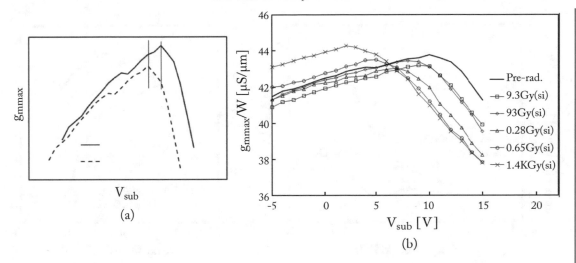

Figure 4.2: (a) Schematic $g_{mmax} - V_{sub}$ characteristics before and after X-ray irradion. (b) $g_{mmax} - V_{sub}$ characteristics for each X-ray dose.

4.3 CAUSE OF *p*-MOSFET DRAIN CURRENT CHANGE

As mentioned above, the possible causes of *p*-MOSFET drain current change are: (i) $|V_{to}|$ increase by the charge in gate oxide; (ii) positive back-bias by the charge in BOX; (iii) local $|V_{to}|$ increase by the charge in sidewall spacer; and (iv) mobility reduction by the generated interface states. To distinguish which is the dominant cause, analysis by Terada's method [47] is used. In Terada's method, measured source to drain resistance R_m including parasitic source drain resistance R_{ext} is given by

$$R_m = \frac{\rho_{ch}}{W_{eff}}L + \left(\delta L \frac{\rho_{ch}}{W_{eff}} + R_{ext}\right),\qquad(4.3)$$

where ρ_{ch} is the channel sheet resistance, W_{eff} is the effective gate width, L is the gate length in design, and δL is the gate length bias from designed to effective gate length. Using different gate length MOSFET set and varying $V_{gs} - V_{to}$, δL and R_{ext} can be extracted. In addition, ρ_{ch} is given by

$$\rho_{ch} = \frac{1}{\mu C_{ox}\left(V_{gs} - V_{to}\right)}\qquad(4.4)$$

and the mobility μ can be calculated from the linear relationship between $1/\rho_{ch}$ and $V_{gs} - V_{to}$. The extracted change in μ, δL, and R_{ext} is shown in Fig. 4.4 as a function of X-ray dose both for *n*- and *p*-MOSFETs. In the case of *n*-MOSFET, only change of μ by the X-ray irradiation is observed whereas no change in δL and R_{ext}. This can agree with previous analysis of *n*-MOSFET radiation degradation. On the other hand, δL and R_{ext} increase by the irradiation is observed in the case

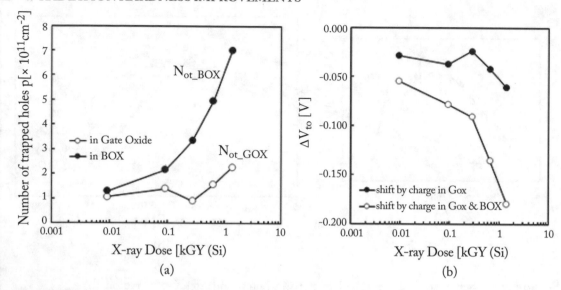

Figure 4.3: (a) Number of trapped holes in gate oxide and BOX as a function of X-ray dose and (b) estimated V_{to} shift from results of (a).

of p-MOSFET, whereas change in μ is small. Therefore, the dominant cause of p-MOSFET drain current change by X-ray irradiation should be related to the gate length modulation and the parasitic resistance increase. With consideration of these phenomena, the cause must be located at the edge of gate. It is suspected that the drain current change is due to the generated positive charge in sidewall spacer and the charge modulates the threshold voltage of the gate edge MOSFET and explained in [48]. In this explanation, the generated number of holes in the sidewall spacer ΔN_{ot_sw} is a function of $\Delta \delta L$ as

$$\Delta N_{ot\_sw} = \frac{\epsilon}{2qL_{eff1}T_{ox\_sw}}\Delta \delta L, \tag{4.5}$$

where L_{eff1} is the effective gate length of the gate edge MOSFET and T_{ox_sw} is the effective sidewall spacer thickness. From calculation of Eq. (4.5) using $\Delta \delta L$, ΔN_{ot_sw} is estimated and compared to the generated number of holes in BOX, as shown in Fig. 4.5. We confirmed the same trend and the gate length modulation is due to the generated positive charge in sidewall spacer.

4.4   IMPROVEMENT OF p-MOSFET RADIATION HARDNESS

The cause of the drain current change of p-MOSFET is the threshold voltage change at the edge of gate due to the generated positive charge in sidewall spacer. To suppress this effect, all channel

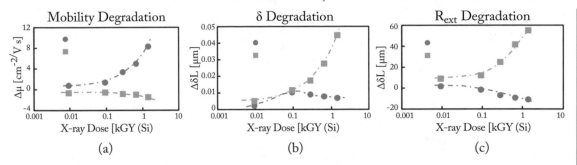

(a)   (b)   (c)

Figure 4.4: Extracted μ, $\delta\Delta L$, and R_{ext} by Terada's method as a function of X-ray dose.

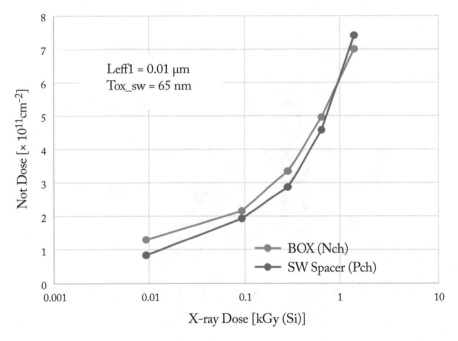

Figure 4.5: Generated number of holes in BOX and sidewall spacer.

of MOSFET must be controlled by gate potential not by the charge in sidewall spacer. Then, the lightly doped drain (LDD) region must be totally overlapped by the gate and higher LDD dose is suspected to improve this effect. Figure 4.6 shows the drain current change ratios for different p-type LDD (PLDD) dose after 112 kGy(Si) X-ray irradiation. It is clear that higher PLDD dose such as 6 times or 10 times improves the ratios from 80% to 20%. This is really great improvement for the radiation hardness.

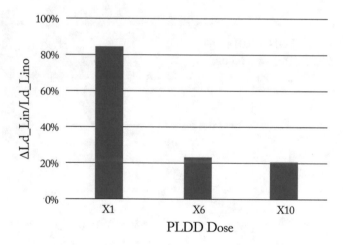

Figure 4.6: Drain current variations ratios for different PLDD dose.

CHAPTER 5

Advanced Process Developments

5.1 DOUBLE SOI

Introducing the BPW layer solves the back-gate issue, but there still remain two issues to realize high-performance SOI detector; crosstalk between circuits and sensors and oxide charge field generated by radiation. The crosstalk induces unwanted signals and makes operation of the detector unstable. The oxide trapped charge alter the electric field in transistor channel, and cause threshold voltage shift and leakage current increase in the transistor.

To solve these issues, Double SOI [41], which has additional silicon and BOX layers in the normal SOI wafer is introduced. Cross section of processed double SOI wafer is shown in Fig. 5.1. The additional middle silicon layer (SOI2) can also take over protecting functions of the BPW layer. Figure 5.2 shows the effectiveness of the middle *Si* to suppress the back-gate effect, and perfect shielding is confirmed. Thus, the BPW can be optimized without considering the back-gate effect [49].

Two kinds of double SOI wafer have been made. Specifications of the wafers are summarized in Table 5.1. The first wafer (D-1) has *p*-type middle *Si* layer of 88 nm thick and *n*-type substrate, and the second one has *n*-type *Si* layer of 150 nm thick and *p*-type substrate. The resistivity of the middle SOI is less than 10 Ohm•cm. Although calculated sheet resistance of the middle-*Si* is very high (< 1 MOhm/square) due to the thin thickness, actual resistance drops to ~30 kOhm/square in the case of D-2 wafer at $V_{SOI2} = 0$ V, as shown in Fig. 5.3. This can be explained by the generation of accumulation layer in the *n*-type silicon by the surrounding oxide. In the *p*-type silicon case (D-1), depletion of the silicon occurs and it shows very high resistance at $V_{SOI2} = 0$ V. Since we will apply negative voltage to the SOI2 layer to compensate oxide charge after irradiation, the resistance of the D-2 will drop further. Sheet resistance of the D-1 also decreases with negative SOI2 voltage due to the formation of inversion layer.

It is important to create enough number of contacts to the middle *Si* layer in the pixel layout to suppress crosstalk between circuit blocks and sensing node. With the middle-*Si* layer and its contacts, crosstalk from a circuit is greatly suppressed, as shown in Fig. 5.4 [50].

When we irradiate X-rays to the SOI chip, transistor threshold voltage will shift to negative direction due to hole trapping in the surrounding oxide. Since the gate oxide is very thin, the major source of the shift is coming from the BOX (see Section 4.2). By applying negative voltage (V_{SOI2}) to the middle *Si*, electric field generated by the trapped hole and/or interface state can

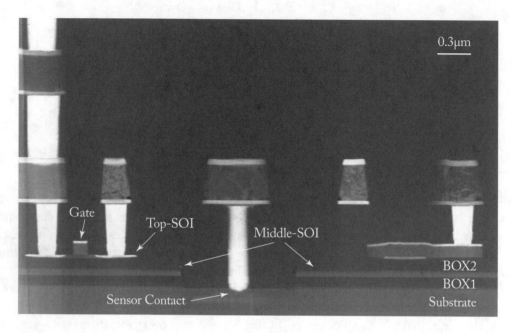

Figure 5.1: Cross sectional view of the double SOI wafer. In addition to top SOI and substrate, a middle SOI layer is added.

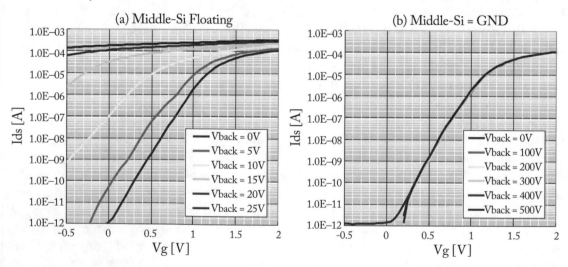

Figure 5.2: Back-gate effect suppression by middle *Si* (@*Vds* = 0.1 V). (a) Middle *Si* is floating. (b) Middle *Si* is connected to ground and the back gate effect is fully suppressed.

Table 5.1: Specification of the double SOI wafers

Layer	D-1	D-2
SOI1	p-type 88 nm, <10 Ohm•cm	p-type 88 nm, <10 Ohm•cm
BOX1	145 nm	145 nm
SOI2	p-type 88 nm, <10 Ohm•cm	n-type 150 nm, <10 Ohm•cm
BOX2	145 nm	145 nm
Subtrate	n-type CZ 725μm, >700 Ohm•cm	p-type CZ, 725μm, >1.0 kOhm•cm

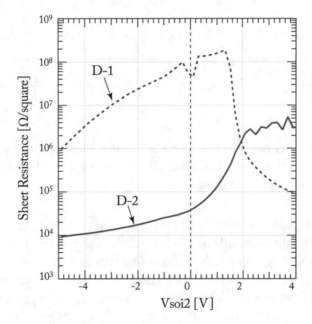

Figure 5.3: Measurement of SOI2-layer sheet resistance vs. SOI2 potential.

be compensated and the threshold voltage will return to almost original value (Fig. 5.5) [51, 52]. The optimum V_{SOI2} is found to show some modest differences among the samples, NMOS and PMOS, source-tied, and body floating.

In addition to the threshold shift, there is some change in drain current as explained in Chapter 4 and this can be fixed by increasing the dose in the LDD region.

5.2   STITCHING

Large area detectors are often required in some experiments, but the mask size of the process is limited. Therefore, stitching technique is developed to make larger format detectors by using

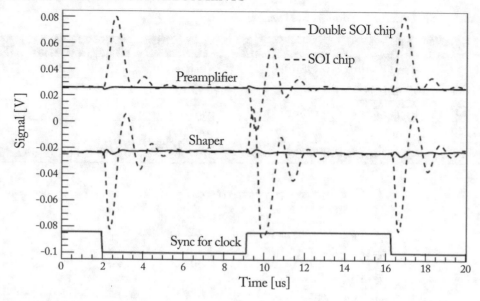

Figure 5.4: Suppression of crosstalk by the double SOI wafer. Pre-amplifier and shaper output signal is observed with running clock signal. Crosstalk from counter appeared in the single SOI wafer is greatly reduced by using the double SOI wafer.

only one mask set. The mask pattern is connected at adjacent shots to cover large chip area. Figure 5.6 shows the stitching method and photographs of the processed wafer. The development is mainly driven by RIKEN group for the Silicon-On-Insulator Photon Imaging Array Sensor (SOPHIAS) detector [53]. The shot is repeated three times then edge pattern is added. The size of the detector is about 3 cm by 6 cm. Major difficulty in making the large sensor is low yield. Any defects will obstacle the perfect detector. Therefore, more loose design rules are used in designing the SOPHIAS detector. The detail of the SOPHIAS detector is explained in Section 6.5.

5.3   BACK-GATE PINNED SOI PIXEL

In order to increase the conversion gain and reduce the back-gate effect, a new detector with charge collection and surface potential pinning structure is proposed (BPSPIX: Back-gate Pinned SOI Pixel) [54]. Figure 5.7 shows the structure and potential profile of the proposed detector.

A charge collector $n+$ and two different buried n-wells, BNW1 and BNW2, are formed in the n-type high-resistivity substrate. $p+$ layer is formed at the backside of the substrate with a back-end process steps. Thanks to the pinned surface potential of the back gate with the BPW, the SOI CMOS circuits are isolated from the photo detector in the substrate and the operation is stabilized. The BPW also acts as a pinning layer to reduce the dark current. The fast charge correction inside the pixel is done by a depleted potential profile created in the substrate. The

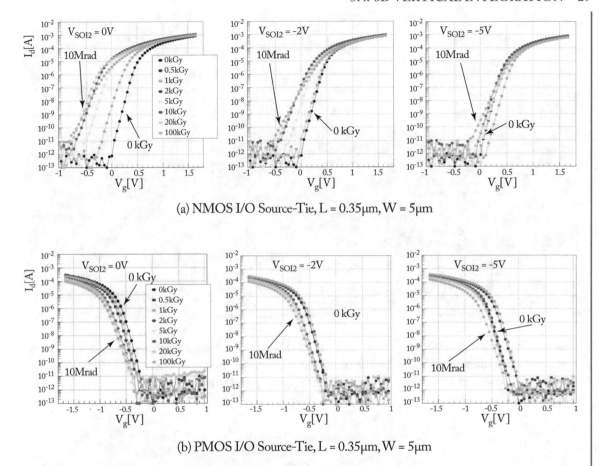

(a) NMOS I/O Source-Tie, L = 0.35μm, W = 5μm

(b) PMOS I/O Source-Tie, L = 0.35μm, W = 5μm

Figure 5.5: Id-Vg curve change with γ-ray irradiation and middle *Si* voltage (VSOI2). (a) NMOS, and (b) PMOS. All terminal was grounded during the irradiation.

BNW2 plays an important role for creating lateral electric field for high-speed charge correction and increasing a potential barrier to holes in the BPW. This allows us to use this detector under a fully depleted condition by applying negative voltage to the backside $p+$ region while preventing the punch-through to the back-gate (BPW region) and injection of holes from the BPW. Since the BNW1 and BNW2 are fully depleted and the capacitance of the charge collector is only due to the $n+$/BPW junction, high charge-to-voltage conversion gain is realized.

5.4   3D VERTICAL INTEGRATION

3D vertical integration is desirable in smart pixel applications, since it can enhance pixel functions without increasing its pixel size. Future pixel detectors such as those used in the International

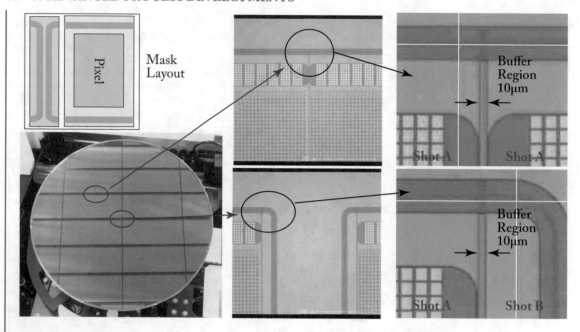

Figure 5.6: Stitching exposure for the SOPHIAS detector. Both edge structure and pixel structure were drawn in a mask. Part of the mask is blocked during exposure. In this detector (SOPHIAS), pixel layout was repeated three times and edge structures are exposed at both ends.

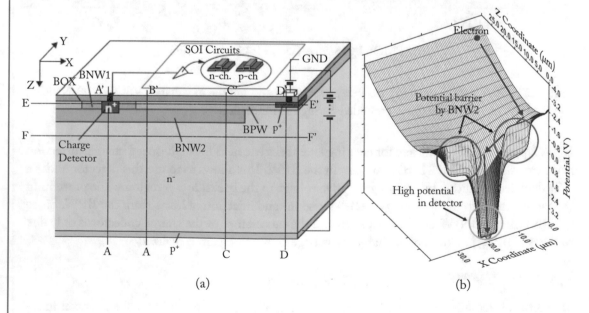

Figure 5.7: Sensor structure with backside depletion.

Linear Collider (ILC) project requires much more transistors to implement memories and data processing logics than available today [26]. If memory and logics are implemented inside the pixel, the pixel size has to be very large ($> 60 \times 60 \ \mu m^2$), while the experiment requires smaller pixel size less than $20 \times 20 \ \mu m^2$. To solve this, development of 3D vertical integration technology is essential [55]. A 3D chip is generally referred to as a chip comprised of two or more tiers of active semiconductor devices that have been thinned, bonded, and interconnected to form a "monolithic" circuit [56]. As the performance improvement, 3D integration is expected to increase space and time resolution, functionality without increasing pixel size. For example, sensors and analog circuits are fabricated in lower tier and digital circuits and I/Os are in upper tier.

Figure 5.8 shows the process flow of the vertical stacking of circuit layers using gold cone-bump connection [57] by T-micro Co. Ltd. [58]. Base SOI chips as upper and lower tiers are fabricated in a SOI wafer. A gold (*Au*) cone bumps are formed by nanoparticle deposition method. The fabricated prototype has 3-μm-diameter *Au* cone-bump connections with adhesive injection. The resistance of the *Au* cone bump obtained from a 2,000 bump daisy chain is less than 1 Ω/bump [59].

5.5   SUPER-STEEP SUBTHRESHOLD SLOPE TRANSISTOR

To get high-sensitivity and low-power characteristic of the sensor, it is better to have a very steep characteristic in Id-Vg curve. By using the Lapis SOI pixel process, super-steep subthreshold slope (SS) transistor is being studied [60, 61]. Normal SOI transistor has SS of ~67 mV/dec, but new structure transistor shows < 6 mV/dec SS (Fig. 5.9). The drain current increases more than two orders with a little change of the gate voltage. Utilization of this characteristic in sensor application is being pursued.

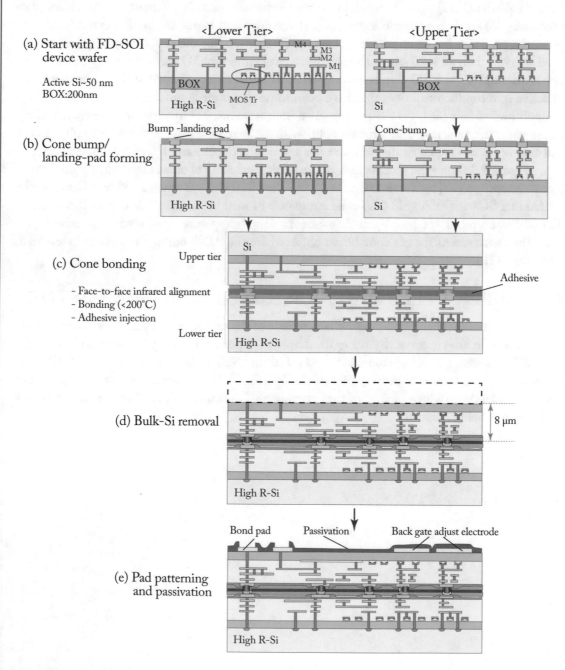

Figure 5.8: Process flow for the stack process using gold cone-bump connection.

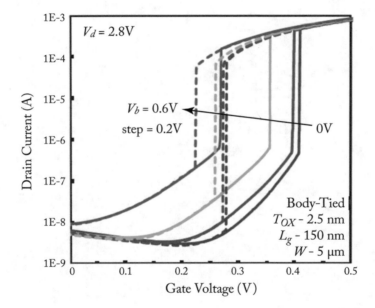

Figure 5.9: An example of the SST transistor characteristics. Hysteresis characteristics are measured depend on the body voltage V_b with the Body-Tie transistor of the core NMOS.

CHAPTER 6

Detector Research and Developments

There have been many kinds of the SOI pixel detectors developed for a variety of applications. In this section, major activities of the applications, developments, and tests are introduced briefly.

The cost of semiconductor process is not cheap, and major part of the cost is mask set. To reduce development cost and have multiple chances for design submission, KEK group has been operating Multi Project Wafer (MPW) runs of the SOI pixel process from 2007 about twice per year. There have been many submissions from all over the world in addition to Japanese institutes. Present mask size is 24.6 mm × 30.8 mm and an example of MPW wafer is show in Fig. 6.1.

Figure 6.1: Wafer fabricated in SOI Pixel Multi Project Wafer run.

6.1   INTEGRATION TYPE PIXEL

Main SOI detectors developed so far are integration-type pixel detectors [62–64]. The generated charge in the sensing region is integrated in a capacitor during a measuring time. Then, the voltage of the capacitor is read out to outside. The first integration-type pixel detector (called INTPIX) with 32×32 arrays of 20 μm square pixel in 2.4 mm square chip had been developed in 2006 [15] defined. A basic schematic and a layout of an INTPIX pixel are shown in Fig. 6.2. The circuit is similar to that of the CMOS optical imager. Many of the integration-type pixels have CDS circuit in each pixel or in column circuit. An example of X-ray CT image taken by INTPIX4 detector is shown in Fig. 6.3. The image is taken with 9.5 keV synchrotron X-ray at the KEK photon factory. The INTPIX4 detector has about 512×832 ($= 426$ k) pixels of 17 μm square. Another example which measures Compton scattered electron track by irradiating high-energy X-rays is shown in Fig. 6.4.

Some of the INTPIX chips were successfully thinned from 50~100 μm to use as a vertex detector in high-energy accelerator experiments [65]. The thinned wafer was diced and the thinning quality is evaluated. There was only a small increase in the leakage current and in the breakdown onset voltage but there was no operation problem. The thinned device was characterized by injecting penetrating infrared laser, and with back illuminated red laser. The bias dependence of the collected charge was as expected, showing that the device thinned to 100 μm was fully depleted at around 90 V. The devices were tested in a high-energy beam, providing a signal-to-noise ratio of 40 at room temperature.

6.2   COUNTING-TYPE PIXEL (CNTPIX)

Capability of containing complex functions in a pixel is one of the fascinating features in the SOI pixel detector. The CNTPIX series [66, 67] is photon counting-type pixel detectors for X-rays, which contains both analog amplifiers and digital logics. A typical schematic of the counting-type pixel is shown in Fig. 6.5. The counting-type pixel counts the number of photon signals that cross the preset threshold. The CNTPIX5 chip is 72×218 arrays of a 64 μm square pixel, and the chip size is 5.0 mm $\times$ 15.4 mm. Each pixel has two discriminators with 3-bits DAC for fine adjustment of threshold. After the analog signal is discriminated, the pulse is counted in 9-bits $\times$ 2 counters. The counters can also be configured as 18-bits or 9-bits $\times$ 8 counters with adjacent 4 pixels. Although basic function is confirmed to work, it suffered crosstalk between sensor and circuit. This can be avoided by using the double SOI wafer (Section 5.1).

6.3   X-RAY DETECTOR FOR ASTROPHYSICS (XRPIX)

Imaging spectrometers for X-ray astronomy are required to have good sensitivity for large range of X-ray energy, adequate radiation hardness, and low-power consumption. X-ray CCDs are widely used at present for their good performance [68, 69]. However, there are some disadvantages in X-ray CCD, such as low readout rate, poor radiation hardness, and difficulty to distinguish non-

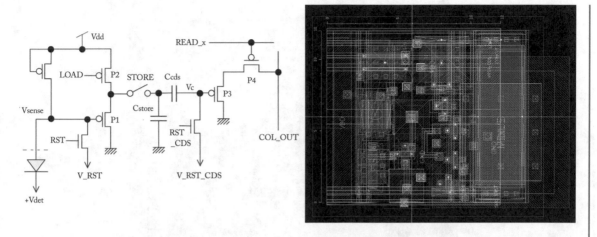

Figure 6.2: (Left) A typical circuit schematic of the integration-type pixel with CDS function. (Right) Layout of INTPIX7 pixel. The size of the pixel is 12 μm × 12 μm.

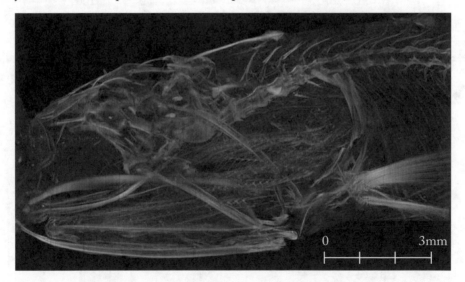

Figure 6.3: An example of X-ray CT image (small fish) taken by the integration-type SOI sensor (INTPIX4) with 9.5 keV synchrotron X-ray. The detector has 500 μm thick FZ sensor and it is operated at detector voltage of 220 V.

X-ray background (NXB). The main component of the NXB is high-energy charged particle of the cosmic ray. Hit rate from the NXB becomes higher than X-ray rate above 10 keV, and it is difficult to remove the NXB in coincidence due to the slow readout rate of a CCD.

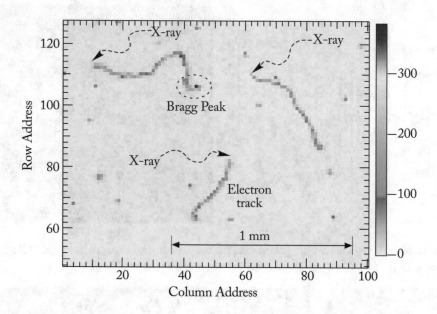

Figure 6.4: Trace of Compton scattered electron from high-energy X-rays irradiation measured with the INTPIX4 detector. Since higher energy deposition (Bragg peak) will occur in the end point of a track, lower energy edge is the collision point of the X-ray.

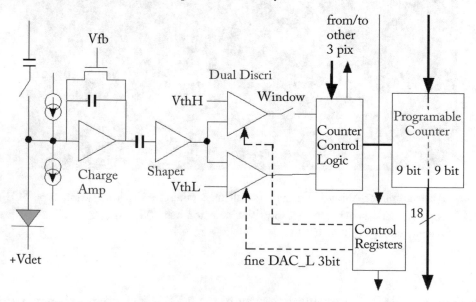

Figure 6.5: Block diagram of a pixel circuit in the CNTPIX5.

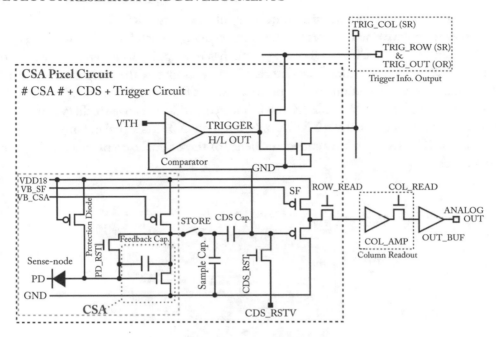

Figure 6.7: XRPIX pixel circuit using charge sensitive amplifier.

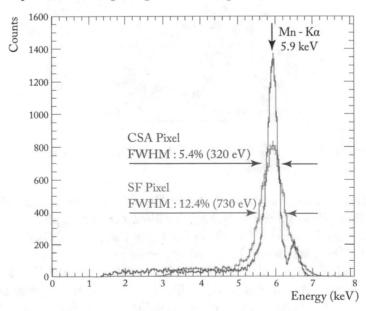

Figure 6.8: Energy spectrum of ^{55}Fe taken with the XRPIX detectors. CSA-type pixel showed better resolution [79].

SOI detector for astronomical satellite X-ray observation (XRPIX) [70–76] has been developed in order to achieve a new X-ray detector system having a much faster readout and lower NXB sensitivity than the X-ray CCDs. Of course, it still has to keep similar spectroscopic performance with the CCD. Basic structure of the detector is same as that of the integration-type detector, but it also has trigger generation and hit position output functions (event-driven readout).

As shown in Fig. 6.6, the XRPIX will be located within an active shield system. In most of the case, X-rays lefts signals in the XRPIX only, but the NXB lefts signal in both the active shield system and the XRPIX. By taking anti-coincidence of these signals, event generated by the NXB can be rejected.

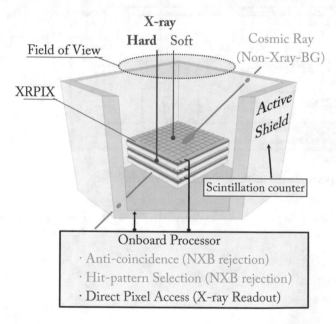

Figure 6.6: Concept of an active shield system with the XRPIX to reduce the non-X-ray background by ant-coincidence with the active shield.

The pixel circuit of the XRPIX is shown in Fig. 6.7. The circuit crosstalk with adjacent pixels is less than 0.5%. Relevant results on backside illumination SOI pixel sensor with different backside process are described in [77]. X-ray event-driven readout mode is successfully operated, but the energy resolution was worse than that of the non-event-driven readout mode due to crosstalk between analog and digital signals [78]. Noise level of the XRPIX is decreasing steadily and approaching the same level of noise with the CCD. One of the big improvements is adopting Charge Sensitive Amplifier (CSA) instead of source-follower amplifier. It shows higher gain and better resolution [79] (Fig. 6.8).

The comparison of XRPIX and X-ray CCD are summarized in Table 6.1. For the NXB, energy range, and time resolution, the XRPIX has better characteristics than those of X-ray CCD.

Although XRPIX have lower energy resolution, the resolution has been improved constantly by optimizing the sensor structure and electronics [80, 81].

Table 6.1: Specification comparison of X-ray CCD and X-ray SOI [74]

Item	X-ray CCD	X0ray SOI	
		Target specification	Achieved specification by 2014
Pixel size	24–30 μm	30–60 μm	30 μm
Chip size	20–30 μm	25 μm	2.4–6 mm
FWHM	130 eV at 6 keV	≤140 eV at 6 keV; ≤230 eV at 22 keV	300 eV at 5.9 keV
Readout noise	3–6 e-rms	≤10 e-rms 3 e-rms (Goal)	33 c-rms
Energy range	0.3 keV - 10 keV	0.3 keV - 40 keV	0.3 keV - 40 keV
Time resolution	A few seconds	10μs per event readout Throughput: 2kHz	10μs per event readout Throughput: 1kHz

6.4 VERTEX DETECTOR FOR CHARGED PARTICLES (PIXOR)

Vertex detector is a high-resolution tracking detector to measure charged particle track in micron accuracy. It plays very important role in recent high-energy accelerator experiments. PIXOR (Pixel OR) is a research for future vertex detectors such as Belle II experiment, which is being developed by Tohoku University and KEK group [82]. The PIXOR readout scheme (Fig. 6.9) is based on a unit of an $N \times N$ pixel matrix, called Super Pixel. An analog signal from each pixelated sensor is divided into two-dimensional directions, and $2N$ signal channels from a small N-by-N pixel matrix are wired-OR as N column and N row channels. Then the signals are processed by a readout circuit in each small matrix and wait for a trigger (Fig. 6.10).

The PIXOR scheme reduces the number of readout channels from N^2 to $2N$ still keeping same position resolution. Although there might be ghost hit if there are more than two hits in a Super Pixel before readout, it can be used in modest hit rate environment. This scheme avoids a deterioration of intrinsic position resolution due to pixel area since only $2N$ circuits are implemented in N^2 pixel area. Thus, the PIXOR detector represents an intermediate solution between pixels and strips. The feature of PIXOR allows good point resolution, low occupancy, low material budget, and on-sensor signal processing at the same time.

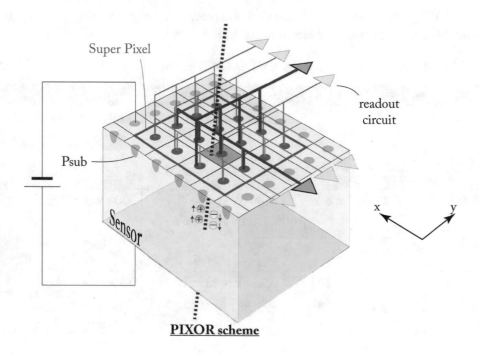

Figure 6.9: Conceptual view of the PIXOR detector showing 4 × 4 Super Pixel case. Actual sensor being developed has 16 × 16 Super Pixel. The figure is showing a charged particle penetrating the green pixel, and the detected signal is divided into X and Y direction. Amplifiers and readout circuit is located above the super pixel in real detector.

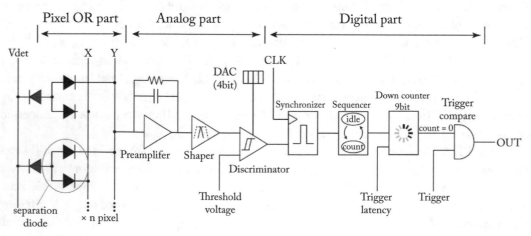

Figure 6.10: Block diagram of the single-channel PIXOR1 circuit.

6.5   XFEL DETECTOR (SOPHIAS)

SOPHIAS is a large dynamic range X-ray image sensor [83] developed by Riken group for X-ray Free-electron Laser Facility (XFEL), SPring-8 Angstrom Compact free electron laser (SACLA) [84]. The target application is the coherent X-ray diffraction imaging experiments. The sensor fabricated from SOI wafer enables monolithic CMOS sensor with fully depleted p-n junction diode as thick as 500 μm. To fabricate larger sensor (64.77×26.73 mm^2) than mask size, stitching technique is developed and used (Section 5.2). Each pixel has high- and low-gain channels to achieve large dynamic range. The different gain is achieved by different value of the input capacitor and different number of sensing nodes (Fig. 6.11).

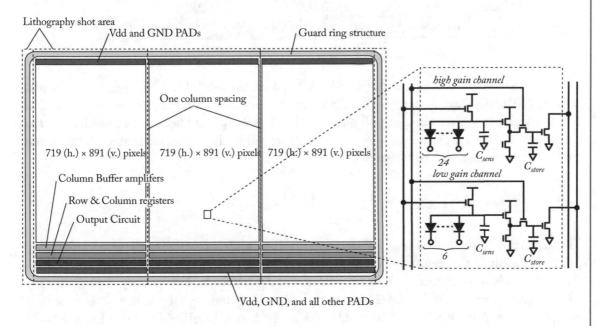

Figure 6.11: Outlook of the SOPHIAS detector and schematics of the pixel.

Table 6.2 compares the specifications of the detectors for SOPHIAS and multi-port Charge-coupled device (MPCCD) that has been developed at SPring-8 and has been used in the SACLA from 2014 [85, 86].

6.6   FERMILAB SOI CMOS DETECTOR (MAMBO)

Monolithic Active Matrix with Binary Counters (MAMBO) [88] is a counting-type detector designed by Fermilab group for detecting and measuring low-energy X-rays from 6–12 keV. Each pixel consists of a charge-sensitive preamplifier, shaper, baseline restorer, window comparator, 4-bit trimming DACs and a 12-bit ripple counter which can be reconfigured as a shift register.

Table 6.2: Sensor specifications of MPCCD and SOPHIAS

	MPCCD (Phase I)	SOPHIAS
Pixel Size (μm)	50	30
Pixel Number	0.53 M	1.9 M
Readout Speed (MHz)	3.3 or 5.4	20
Frame Rate (Hz)	60	60
Sensitive Layer Thickness (μm)	50	500
Noise (e-rms)	100–250	200
Peak Signal (Me-/200 μm²)	16–20 Me-	185 Me- [87]

A nested well structure with (Section 3.2.3) a buried n-Well (BNW) and deeper buried p-wells (BP1, BP2) are used to electrically isolate the detector from the electronics as shown in Fig. 6.13. The BNW acts as a ground plane to electrical signals and behaves as a shield, and the buried p-well layers act as a sensing node. Preliminary tests indicate that the nested well structure successfully shields the detector and electronics.

6.7   LBNL SOI-IMAGER

The Lawrence Berkeley National Laboratory (LBNL) group is developing depleted monolithic pixel sensors in SOI technology [89, 90]. The purpose is biased toward technology demonstration with high-momentum particles on analog and digital pixels. Two prototype pixel sensors, LDRD-SOI-1 and LDRD-SOI-2, featuring both analog and digital pixels have been subsequently designed at LBNL and fabricated in KEK MPW run. LDRD-SOI-1 and LDRD-SOI-2 are fabricated in a 0.15 μm and 0.20 μm process, respectively. A S/N ratio of 20 is found for high-energy electron hits, with single pixel noise figures of 30 electrons, for a readout frequency up to 50 MHz.

The SOImager-2 sensor which has a matrix of 256 × 256 pixels arrayed on a 13.75 μm pitch is also designed. The chip implements pixel cells of different designs with various combinations of floating p-type guard rings and the BPW. This sensor has been successfully tested with high momentum particles at the CERN SPS and achieved very good resolution, as shown in Fig. 6.14 [91]. The detector is also thinned to be 73 ± 2 μm [77]. The measured quantum efficiency curve is in agreement to that for a sensor with a 0.6 ± 0.2 μm back-plane passive window, in units of equivalent Si thickness.

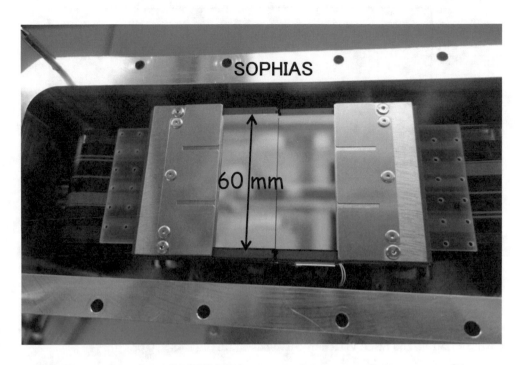

Figure 6.12: Photograph of the SOPHIAS detector and overview of the camera. (Courtesy of T. Kudo [87].)

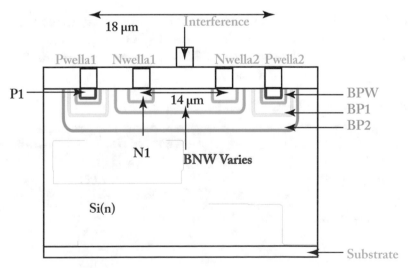

Figure 6.13: Pixel with nested well structure in the MAMBO V detector.

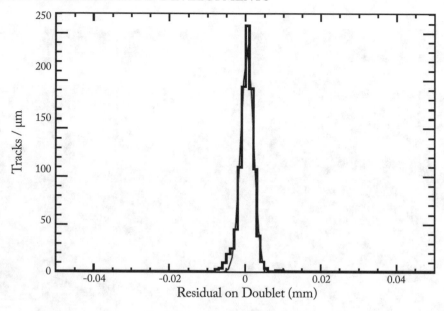

Figure 6.14: Residual distribution for 200 GeV pion track extrapolated on the second plane of the doublet at $V_d = 50$ V. The fitted Gaussian curve has width of (1.48 ± 0.06) μm [91].

6.8   ULTRA-LOW TEMPERATURE APPLICATIONS

Researchers in JAXA/ISAS have interest in using SOI devices in cryogenic temperature. They developed CMOS amplifiers and switches in the LAPIS SOI process, and confirmed to work without problem in low temperature where bulk CMOS devices cannot work [92]. Typical performance of the amplifier is shown in Table 6.3.

Table 6.3: Performance of the FD-SOI CMOS amplifier at 4.2 K

	Design	Measurement
Operating temperature	4 K	4–300 K
Open loop gain	>1,000	7,000
Output voltage swing	>1 V	1.3 V
Input referred noise at 1 Hz	14–20 μV/$\sqrt{\text{Hz}}$	19 μV/$\sqrt{\text{Hz}}$
Power consumption	1.1 μW	1.3 μW
Input offset voltage	0 mV	2 mV
Variation of input offset voltage	0 mV	4.2 mV (1σ)
Leak current of reset switch	0.1 fA	0.1 fA

After the success of the cryogenic operation, University of Tsukuba group started a project to build Superconducting Tunnel Junction (STJ) devices on top of processed SOI wafer [93]. The extraction of STJ signals from below 1 K to room temperature is always annoying issue. By processing the STJ device on processed SOI wafer and make good connections between them, number of the connection can be greatly reduced and S/N can be improved.

Characteristics of SOI transistors at below 1 K is successfully measured and shown in Fig. 6.15 [94]. Both NMOS and PMOS still show good transistor characteristics. Nb/Al STJ device was successfully build on top of the SOI wafer and output signal is observed by illuminating laser light (465 nm) to the device (Fig. 6.16).

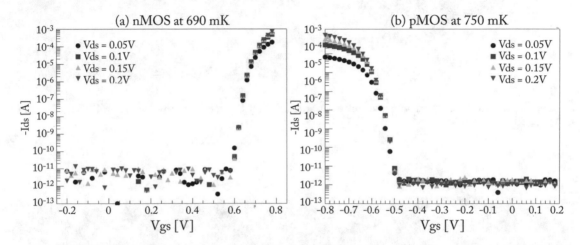

Figure 6.15: Ids-Vgs characteristics of NMOS and PMOS transistors at 690 mK and 750 mK, respectively.

6.9   READOUT BOARD (SEABAS)

To test manufactured chip quickly, SEABAS (SOI EvAluation BoArd with Sitcp) read out board was developed by KEK group. The board contains two FPGAs one for controlling the SOI pixel chip and the other is for transferring the data through Ethernet (called SiTCP) [21, 95].

Figure 6.17 shows photograph of the SEABAS2 (second generation board) and the INTPIX4 sub-board. By using the SEABAS2 board, we could take ~90 frame/s with INTPIX4 detector (425 k pixels).

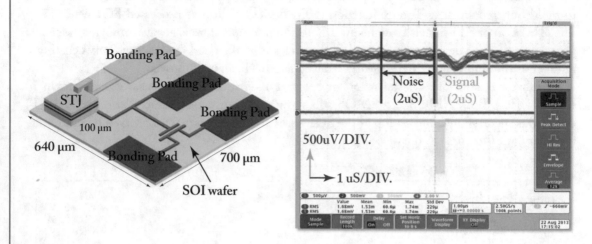

Figure 6.16: (Left) Test structure of an element. (Right) Signal from the Nb/Al-STJ built on the SOI wafer.

Figure 6.17: Photograph of the SEABAS2 board and the INTPIX4 sub-board.

CHAPTER 7

Summary

The SOI pixel detector is a new kind of monolithic radiation image detector which includes both radiation sensing part and readout circuits. There are many superior features in the SOI device compared to bulk CMOS devices such as high speed, low power, lower capacitance, small area, low SEE cross section, etc. Since the bonded SOI wafer has two active silicon layers, it has the ideal structure to build a monolithic radiation detector that has a thick sensing region.

There have been several difficulties in implementing the sensor and circuit in very close region such as the back-gate effect and TID sensitivity. Many efforts to overcome these difficulties were done and new techniques such as buried p-well (BPW), nested well, double-SOI wafer, and 3D integration have been developed. In addition, steady improvements are also being pursued such as use of high-resistivity FZ wafer and stitching technique for large area sensor.

Many projects have started to develop application-specific imaging detectors by using MPW run operated by KEK using the 0.2 μm FD-SOI pixel process. For different application fields and research needs, the ongoing projects of SOI pixel sensors have their own focuses, such as radiation hardness for high-energy physics, large dynamic range and large imaging area for XFEL application, energy resolution and trigger capability for astrophysics X-ray sensors, thin backside window for low energy X-ray detection, and so on.

New techniques are still being developed and this will expand the application field and bring significant improvements. Among these, critical issues are radiation hardness and effective shielding between the electronics circuit and the sensor. Understanding of radiation effects to MOS transistors progressed rapidly and solutions are foreseen. More than a few hundreds kGy(Si) radiation hardness of the SOI detector can be reached in the near future. The newly developed double SOI technique solves the above issues. With biasing the middle Si layer, radiation-induced charge trap in the BOX can be compensated. Middle Si layer of the double SOI also works as a good shield between sensors and electronics, so many smart detectors, which have both analog and complex digital circuits with in a pixel, will be developed without annoying crosstalk.

Furthermore, new sensor structures, such as Back-gate Pinned SOI Pixel and avalanche photo diode, are being proposed. These new structures manipulate several kinds of buried layers and aim to ultra-low noise and high sensitivity detectors.

3D integration is expected to play an important role in future high-performance pixel detector. Implementing complex function inside the pixel is one of the future endeavors, and it can help to improve the performance of many SOI detectors such as vertex detector. 3D integration technology can be used to separate digital and analog circuits, allowing the sensors and analog circuits to be shielded from digital noise and offering separate optimization of sensor, analog, and digital layers.

The SOI pixel sensor was originally developed for high-energy physics. But it also has good application prospects in X-ray imaging, such as synchrotron radiation experiments, medical applications, inspection of contamination, and metal fatigue damage. It also has potential applications for visible light, infrared light, and electron/charged particles/neutron detections. SOI pixel detectors have many inherent advantages and the research is in a phase of rapid development and collaboration.

Bibliography

[1] N. Wermes, Pixel detectors for charged particles, *Nucl. Instr. Meth. A*, vol. 604, no. 1–2, pp. 370–379, June 2009. DOI: 10.1016/j.nima.2009.01.098. 1

[2] E. Schioppa, et al., Prospect for spectral CT with Medipix detectors, *Proc. Sci.*, PoS(TIPP2014)246. 1

[3] M. Pohl, Particle detection technology for space-borne astroparticle experiments, *Proc. Sci.*, PoS(TIPP2014)013. 1

[4] R. Plackett, et al., Current status of the Medipix2, Timepix, Medipix3 and Timepix2 pixel readout chips, *Proc. Sci.*, PoS(VERTEX2010)030. 1

[5] P. Kraft, et al., Characterization and calibration of PILATUS detectors, *IEEE Trans. Nucl. Sci.*, vol. 56, no. 3, pp. 758–764, 2009. DOI: 10.1109/tns.2008.2009448. 1

[6] R. Turchetta, J. D. Berst, B. Casadei, G. Claus, C. Colledani, W. Dulinski, Y. Hu, D. Husson, J. P. Le Normand, and J. L. Riester, A monolithic active pixel sensor for charged particle tracking and imaging using standard VLSI CMOS technology, *Nucl. Instr. Meth. A*, vol. 458, no. 3, pp. 677–689, 2001. DOI: 10.1016/s0168-9002(00)00893-7. 1

[7] J.-P. Colinge, An overview of CMOS-SOI technology and its potential use in particle detection system, *Nucl. Instr. Meth. A*, vol. 305, pp. 615–619, 1991. DOI: 10.1016/0168-9002(91)90164-l. 1

[8] K. Izumi, M. Doken, H. Ariyoshi, C.M.O.S. devices fabricated on buried SiO_2 layers formed by oxygen implantation into silicon, *IEEE Electron. Lett.*, vol. 14, no. 18, pp. 593–594, 1978. DOI: 10.1049/el:19780397. 1

[9] B. Dierickx, et al., Integration of CMOS-electronics and particle detector diodes in high-resistivity silicon-on-insulator wafers, *IEEE Trans. Nucl. Sci.*, vol. 40, pp. 753–758, 1993. DOI: 10.1109/23.256656. 1

[10] F. X. Pengg, Monolithic silicon pixel detectors in SOI technology, Ph.D. dissertation, Institute for Semiconductor Physics, University of Linz, Austria, 1996. http://inspirehep.net/record/887486/files/CM-P00041376.pdf?version=1

[11] C. Xu, W. Zhang, and M. Chan, A low voltage hybrid bulk/SOI CMOS active pixel image sensor, *IEEE Electron Device Lett.*, vol. 22, no. 5, pp. 248–250, 2001. DOI: 10.1109/55.919244. 1

[12] J. Marczewski, et al., Development of a monolithic active pixel sensor using SOI technology with a thick device layer, *IEEE Trans. Nucl. Sci.*, vol. 57, pp. 381–386, 2010. DOI: 10.1109/tns.2009.2036435. 1

[13] A. Bulgheroni et al., Monolithic active pixel detector realized in silicon on insulator technology, *Nucl. Instr. Meth. A*, vol. 535, no. 1–2, pp. 398-403, 2004. DOI: 10.1016/j.nima.2004.07.160.

[14] E. Cortina, et al., TRAPPISTe pixel sensor with 2 μm SOI technology, *Nucl. Instr. Meth. A*, vol. 633, pp. S19–S21, May 2011. DOI: 10.1016/j.nima.2010.06.109. 1

[15] Y. Arai, et al., First results of 0.15 μm CMOS SOI pixel detector, *SNIC Symposium*, Stanford, California, SLAC-PUB-12079, 2006. SLAC Electronic Conference Proceedings Archive (SLAC-R-842, eConf:C0604032) PSN-0016. http://www.slac.stanford.edu/econf/C0604032/papers/0016.PDF 2, 11, 36

[16] T. Ohno, Y. Kado, M. Harada, and T. Tsuchiya, Experimental 0.25-μm-gate fully depleted CMOS/SIMOX process using a new two-step LOCOS isolation technique, *IEEE Trans. Electron Dev.*, vol. 42, no. 8, pp. 1481–1486, 1995. DOI: 10.1109/16.398663. 2

[17] T. Ichimori and N. Hirashita, Fully-depleted SOI CMOS FETs with the fully-silicided source/drain structure, *IEEE Trans. Elec. Dev.*, vol. 49, no. 12, pp. 2296–2300, 2002. DOI: 10.1109/ted.2002.807443. 2

[18] M. Bruel, U.S. Patent No. 5,374,564 (December 20, 1994). 2

[19] G. K. Celler and Sortin Cristoloveanu, Frontiers of silicon-on-insulator, *J. Appl. Phys.*, vol. 93, no. 9, pp. 4955–4978, 2003. DOI: 10.1063/1.1558223.

[20] SOITEC, http://www.soitec.com/ 2, 11

[21] Y. Arai, et al., Developments of SOI monolithic pixel detectors, *Nucl. Instr. Meth. A*, vol. 623, no. 1, pp. 186–188, Nov. 2010, doi:10.1016/j.nima.2010.02.190. DOI: 10.1016/j.nima.2010.02.190. 2, 11, 47

[22] Y. Arai, et al., Development of SOI pixel process technology, *Nucl. Instr. Meth. A*, vol. 636, no. 1, pp. S31–S36, Apr. 2011. DOI: 10.1016/j.nima.2010.04.081. 2

[23] L. Soung et al., Test of the TRAPPISTe monolithic detector system, *Nucl. Instr. Meth. A*, vol. 731, pp. 141–145, Dec. 2013. DOI: 10.1016/j.nima.2013.03.052. 2

[24] SOI pixel collaboration website, http://rd.kek.jp/project/soi/index.html

[25] T. Miyoshi, et al., Recent progress of pixel detector R&D based on SOI technology, *Phys. Procedia*, vol. 37, pp. 1039–1045, Jan. 2012. DOI: 10.1016/j.phpro.2012.02.450. 2

[26] Y. Arai and M. Motoyoshi, Application of 3D stacking technology to SOI radiation image sensor, *IEEE Electr. Des. Adv. Pack. Syst. (EDAPS)*, IEEE Part Number: CFP13EDP-USB, SS1-2, pp. 5-8, Dec. 2013. DOI: 10.1109/edaps.2013.6724443. 2, 31

[27] Lan Hao, et al., A review of SOI monolithic active pixel sensors for radiation detection applications, *IEEE Sensors J.*, 2015, DOI: 10.1109/JSEN.2015.2389271. DOI: 10.1109/jsen.2015.2389271. 2

[28] Y. Arai, et al., SOI pixel developments in a 0.15µm technology, *IEEE Nuclear Science Symposium Conference Record*, vol. 2, pp. 1040–1046, 2007. DOI: 10.1109/nss-mic.2007.4437189. 5

[29] T. Tsuboyama, et al., R&D of a pixel sensor based on 0.15 µm fully depleted SOI technology, *Nucl. Instr. Methods Phys. Res. A*, vol. 582, no. 3, pp. 861–865, Dec. 2007.

[30] S. Glab, et al., Characterization of transistors fabricated in evolving LAPIS semiconductor silicon-on-insulator 0.2µm technology, *Proc. 20th Int. Conf. Mixed Design of Integr. Circuits Syst. (MIXDES)*, pp. 360–364, Gdynia, 2013. 5

[31] K. Hirose, et al., SEU resistance in advanced SOI-SRAMs fabricated by commercial technology using a rad-hard circuit design, *IEEE Trans. Nucl. Sci.*, vol. 49, no. 6, pp. 2965–2968, 2002. DOI: 10.1109/tns.2002.805978. 7

[32] S. Mattiazzo, et al., Total dose effects on a FD-SOI technology for monolithic pixel sensors, *IEEE Trans. Nucl. Sci.*, vol. 57, pp. 2135–2141, 2010. DOI: 10.1109/tns.2009.2038378. 7

[33] Y. Onuki, et al., SOI detector developments, in *Proc. of the 20th Anniversary International Workshop on Vertex Detectors, Proc. Sci. PoS (Vertex)043*, Austria, June 2011. 7

[34] J. P. Colinge, *Silicon-on-Insulator Technology Materials to VLSI*, 3rd ed., Springer Science+Business Media, Inc., 2004. DOI: 10.1007/978-1-4757-2611-4. 9

[35] M. Watanabe and A. Tooi, Formation of the SiO2 films by oxygen-ion bombardment, *Jpn. J. Appl. Phys.*, vol. 5, p. 737, 1966. DOI: 10.1143/jjap.5.737. 10

[36] A. Wittkower et al., SMART-CUT® technology for SOI: A new volume application for ion implantation, *Conf. Ion Implantation Tech.*, pp. 269–272, 2000. DOI: 10.1109/iit.2000.924141. 10

[37] SOITEC. http://www.soitec.com/ 10

[38] I. Kurachi, Review of FD-SOI technology and its application to X-ray pixel sensors, *VIPS2010-Workshop on Vertically Integrated Pixel Sensors*, pp. 24–02, Pavia, Italy, 2010. 10, 11

[39] T. Ichimori and N. Hirashita, Co salicide technology for Sub-0.15 um FD-SOI and beyond: Super-flat silicide and fully-silicided source/drain structure, *IEEE Intl. SOI Conf.*, pp. 72–73, 2000. DOI: 10.1109/soi.2000.892775. 12

[40] F. Ichikawa et al., Fully depleted SOI process and device technology for digital and RF applications, *Solid-State Electron.*, vol. 48, no. 6, pp. 999–1006, 2004. DOI: 10.1016/j.sse.2003.12.028. 12

[41] M. Okihara et al., Progress of FD-SOI technology for monolithic pixel detectors, in *Proc. IEEE Nucl. Sci. Symp. Med. Imag. Conf. Rec. (NSS/MIC)*, pp. 471–474, 2012. DOI: 10.1109/nssmic.2012.6551151. 12, 25

[42] K. Hara, et al., Radiation resistance of SOI pixel devices fabricated with OKI 0.15 μm FD-SOI technology, *IEEE Trans. Nucl. Sci.*, vol. 56, no. 5, pp. 2896–2904, Oct. 2009. DOI: 10.1109/tns.2009.2028573. 16

[43] M. Kochiyama, et al., Radiation effects in silicon-on-insulator transistors with back-gate control method fabricated with OKI Semiconductor 0.20μm FD-SOI technology, *Nucl. Instr. Meth. A*, vol. 636, no. 1, pp. S62–S67, Apr. 2011. DOI: 10.1016/j.nima.2010.04.086. 16

[44] F. F. Khalid, G. W. Deptuch, A. Shenai, and R. J. Yarema, Monolithic active pixel matrix with binary counters (MAMBO) ASIC, in *Proc. Nuclear Science Symposium Conference Record (NSS/MIC)*, pp. 1544–1550, 2010. DOI: 10.1109/nssmic.2010.5874035. 17

[45] J. R. Schwank et al., Radiation effects in SOI technology, *IEEE Trans. Nucl. Sci.*, vol. 50, no. 3, pp. 522–538, 2003. DOI: 10.1109/tns.2003.812930. 19

[46] J. R.. Schwank et al., Radiation effects in MOS oxide, *IEEE Trans. Nucl. Sci.*, vol. 55, no. 4, pp. 1833–1853, Aug. 2008. DOI: 10.1109/tns.2008.2001040. 19

[47] K. Terada and H. Muta, A new method to determine effective MOSFET channel length, *Jpn. J. Appl. Phys.*, vol. 18, no. 5, pp. 953–959, 1979. DOI: 10.1143/jjap.18.953. 21

[48] I. Kurachi, et al., Analysis of effective gate length modulation by x-ray irradiation for fully depleted SOI p-MOSFETs, *IEEE Trans. Electron Devices*, vol. 62, no. 8, pp. 2371–2376, 2015. DOI: 10.1109/ted.2015.2443797. 22

[49] Y. Arai, Progress of SOI pixel process. Presented at PIXEL2012. https://indico.cer n.ch/event/137337/session/3/contribution/110 25

[50] Y. Lu, A study on the shielding mechanisms of SOI pixel detector, *International Workshop on SOI Detector (SOIPIX 2015)*. https://kds.kek.jp/indico/event/17786/sessio n/3/contribution/19/material/slides/0.pdf 25

[51] Y. Arai, Progress on silicon-on-insulator monolithic pixel process, in *Proc. 22nd International Workshop on Vertex Detectors (Vertex)*, Lake Starnberg, Germany, Sep. 2013. 27

[52] S. Honda, et al., Total ionization damage effects in double silicon-on-insulator devices, in *Proc. Nuclear Science Symposium and Medical Imaging Conference (NSS/MIC)*, pp. 1–7, Seoul, Korea, 2013. DOI: 10.1109/nssmic.2013.6829541. 27

[53] T. Hatsui, Development and deployment status of x-ray 2D detector for SACLA. Presented at PIXEL2012. https://indico.cern.ch/event/137337/session/4/contribution/112 28

[54] H. Kamehama, et al., Fully depleted SOI pixel photo detectors with backgate surface potential pinning, *International Image Sensor Workshop (IISW)*, 2015. http://www.imagesensors.org/PastSessions/Session_3/3--05_KAMEHAMA.pdf 28

[55] M. Motoyoshi, et al., Stacked SOI pixel detector using versatile fine pitch μ-bump technology, in *3D Systems Integration Conference (3DIC), 2011 IEEE International*, pp. 1–4, 2012. DOI: 10.1109/3dic.2012.6262959. 31

[56] R. Yarema, 3D circuit integration for vertex and other detectors, in *Proc. 16th International Workshop on Vertex Detectors*, Lake Placid, NY, Sep. 2007. 31

[57] M. Motoyoshi, et al., 3D integration technology for sensor application using less than 5μm-pitch gold cone-bump connection, *J. of Instrumentation*, 2015. JINST 10 C03004, doi:10.1088/1748–0221/10/03/C03004. DOI: 10.1088/1748-0221/10/03/c03004. 31

[58] Tohoku-Micro Tec (T-Micro) Co. Ltd. http://www.t-microtec.com 31

[59] T-Micro Co. Ltd., private communication. 31

[60] T. Mori and J. Ida, Possibility of SOI based super steep subthreshold slope MOSFET for ultra low voltage application, *IEEE SOI-3D-Subthreshold Microelectronics Technology Unified Conference (S3S)*, pp. 1–2, 2013. DOI: 10.1109/S3S.2013.6716550. DOI: 10.1109/s3s.2013.6716550. 31

[61] J. Ida, et al., Super steep subthreshold slope PN-body tied SOI FET with ultra low drain voltage down to 0.1 V, to be presented at IEDM 2015. DOI: 10.1109/iedm.2015.7409761. 31

[62] T. Miyoshi, et al. Monolithic pixel detectors with 0.2 μm FD-SOI pixel process technology, *Nucl. Instr. Meth. A*, vol. 732, pp. 530–534, Dec. 2013. DOI: 10.1016/j.nima.2013.06.029. 36

[63] M. I. Ahmed, et al., Measurement results of DIPIX pixel sensor developed in SOI technology, *Nucl. Instr. Meth. A*, vol. 718, pp. 274–278, Aug. 2013. DOI: 10.1016/j.nima.2012.10.099.

[64] M. I. Ahmed, et al., Prototype pixel detector in the SOI technology, *J. Instr.*, vol. 9, no. 02, pp. C02010–C02010, Feb. 2014. DOI: 10.1088/1748-0221/9/02/c02010. 36

[65] K. Shinsho, et al., Evaluation of monolithic silicon-on-insulator pixel devices thinned to 100 μm, *Conference Record of IEEE Nuclear Science Symposium (NSS/MIC)*, pp. 646–649, 2010. DOI: 10.1109/nssmic.2010.5873838. 36

[66] K. Hara, et al., Development of INTPIX and CNTPIX silicon-on-insulator monolithic pixel devices, in *19th International Workshop on Vertex Detectors, Proc. of Sci. PoS(VERTEX 2010)033*, Loch Lomond, Scotland, UK, 2010. 36

[67] T. Miyoshi, et al., Performance study of SOI monolithic pixel detectors for X-ray application, *Nucl. Instr. Methods Phys. Res. Sect.*, vol. 636, no. 1, pp. S237–S241, Apr. 2011. DOI: 10.1016/j.nima.2010.04.117. 36

[68] G. P. Garmire, et al., Advanced CCD imaging spectrometer (ACIS) instrument on the Chandra X-ray observatory, in *Proc. SPIE*, vol. 4851, pp. 28–44, 2003. DOI: 10.1117/12.461599. 36

[69] K. Koyama, et al., X-ray imaging spectrometer (XIS) on board suzaku, *Pub. Astron. Soc.*, vol. 59, pp. S23–S33, Japan, 2007. DOI: 10.1117/12.461313. 36

[70] S. Nakashima, et al., Development and characterization of the latest X-ray SOI pixel sensor for a future astronomical mission, *Nucl. Instr. Meth. A*, vol. 731, pp. 74–78, Dec. 2013. DOI: 10.1016/j.nima.2013.04.063. 39

[71] S. G. Ryu, et al., First performance evaluation of an x-ray SOI pixel sensor for imaging spectroscopy and intra-pixel trigger, *IEEE Trans. Nucl. Sci.*, vol. 58, no. 5, pp. 2528–2536, Oct. 2011. DOI: 10.1109/tns.2011.2160970.

[72] A. Takeda, et al., Design and evaluation of an SOI pixel sensor for trigger-driven x-ray readout, *IEEE Trans. Nucl. Sci.*, vol. 60, no. 2, pp. 586–591, Apr. 2013. DOI: 10.1109/tns.2012.2225072.

[73] S. G. Ryu, et al., Design and development of trigger-driven readout with x-ray SOI pixel sensor, *IEEE Nucl. Sci. Symp. Medical Imaging Conf. (NSS/MIC)*, pp. 1197–1200, Conference Record, NP3 M-90, Valencia, 2011. DOI: 10.1109/nssmic.2011.6154601.

[74] T. G. Tsuru, et al., Development and performance of Kyoto's x-ray astronomical SOI pixel (SOIPIX) sensor, in *Proc. SPIE 9144, Space Telescopes and Instrumentation: Ultraviolet to Gamma Ray*, p. 914412, Montreal, Quebec, Canada, Aug. 2014. DOI: 10.1117/12.2057158. 41

[75] S. Nakashima, et al., Progress in development of monolithic active pixel detector for x-ray astronomy with SOI CMOS technology, *Phys. Procedia*, vol. 37, pp. 1373–1380, 2012. DOI: 10.1016/j.phpro.2012.04.100.

[76] S. G. Ryu, et al., Tests with soft x-rays of an improved monolithic SOI active pixel sensor, *IEEE Trans. Nucl. Sci.*, vol. 60, no. 1, pp. 465–469, Feb. 2013. DOI: 10.1109/tns.2012.2231880. 39

[77] M. Battaglia, et al., Characterisation of a thin fully depleted SOI pixel sensor with soft x-ray radiation, *Nucl. Instr. Meth. A*, vol. 674, pp. 51–54, May 2012. 39, 44

[78] A. Takeda, Development and evaluation of event-driven SOI pixel detector for x-ray astronomy, *Proc. Sci.*, POS (TIPP2014) 138, 2014. 39

[79] A. Takeda, et al., Improvement of spectroscopic performance using a charge-sensitive amplifier circuit for an x-ray astronomical SOI pixel detector, PIXEL2014, JINST, 10 C06005, 2015, doi:10.1088/1748–0221/10/06/C06005. DOI: 10.1088/1748-0221/10/06/c06005. 39, 40

[80] H. Matsumura, Improving charge-collection efficiency of SOI pixel sensors for x-ray astronomy, *Nucl. Instr. Meth. A*, vol. 794, pp. 255–259, 2015. doi:10.1016/j.nima.2015.05.008. DOI: 10.1016/j.nima.2015.05.008. 41

[81] H. Matsumura, Investigation of charge-collection efficiency of Kyoto's x-ray astronomical SOI pixel sensors, XRPIX, *Nucl. Instr. Meth. A*, vol. 765, pp. 183–186, 2014. DOI: 10.1016/j.nima.2014.05.025. 41

[82] Y. Ono, et al., Development of the pixel OR SOI detector for high energy physics experiments, *Nucl. Instr. Meth. A*, vol. 731, pp. 266–269, Dec. 2013. DOI: 10.1016/j.nima.2013.06.044. 41

[83] T. Hatsui, et al., A direct-detection x-ray CMOS image sensor with 500 μm thick high resistivity silicon, in *Proc. International Image Sensor Workshop*, Art. Num. 3.05, Utah, 2013. http://www.imagesensors.org/ 43

[84] C. Saji, et al., Evaluation of data-acquisition front ends for handling high-bandwidth data from x-ray 2D detectors: A feasibility study, *Nucl. Instr. Meth. A*, vol. 731, pp. 229–233, Dec. 2013. DOI: 10.1016/j.nima.2013.05.019. 43

[85] T. Hatsui and H. Graafsma, X-ray imaging detectors for synchrotron and XFEL sources, *IUCrJ*, vol. 2(3), pp. 371–383, 2015. DOI: 10.1107/s205225251500010x. 43

[86] T. Hatsui, Developments of x-ray imaging detectors at SACLA/SPring-8: Current status and future outlook, *Synchrotron Radiation News*, vol. 27(4), pp. 20–23, 2014. DOI: 10.1080/08940886.2014.930805. 43

[87] T. Kudo of JASRI, private communication. 45

[88] F. Fahim, et al., Monolithic active pixel matrix with binary counters ASIC with nested wells, *J. Instr.*, vol. 8, no. 4, p. C04008, Apr. 2013. DOI: 10.1088/1748-0221/8/04/c04008. 43

[89] M. Battaglia, et al., Monolithic pixel sensors in deep-submicron SOI technology, *J. Instr.*, vol. 4, no. 4, p. P04007, Apr. 2009. DOI: 10.1088/1748-0221/4/04/p04007. 44

[90] M. Battaglia, et al., Monolithic pixel sensors in deep-submicron SOI technology with analog and digital pixels, *Nucl. Instr. Meth. A*, vol. 604, no. 1–2, pp. 380–384, Jun. 2009. DOI: 10.1088/1748-0221/4/04/p04007. 44

[91] M. Battaglia, et al., Characterisation of a pixel sensor in 0.2 μm SOI technology for charged particle tracking, *Nucl. Instr. Meth. A*, vol. 654, no. 1, pp. 258–265, Oct. 2011. DOI: 10.1016/j.nima.2011.05.081. 44, 46

[92] T. Wada, et al., Development of low power cryogenic readout integrated circuits using fully-depleted-silicon-on-insulator CMOS technology for far-infrared image sensors, *J. Low Temp. Phys*, 167, pp. 602–608, 2012. DOI 10.1007/s10909-012-0461-6. DOI: 10.1007/s10909-012-0461-6. 46

[93] Y. Takeuchi, et al., Development of superconducting tunnel junction detectors as a far-infrared single photon detector for neutrino decay search, *Proc. Sci.*, PoS(TIPP2014)155. DOI: 10.1109/i2mtc.2015.7151327. 47

[94] K. Kasahara, et al., Development of superconducting tunnel junction photon detector on SOI preamplifier board to search for radiative decays of cosmic background neutrino, *Proc. of Sci.*, PoS(TIPP)074, 2014. 47

[95] SEABAS. http://rd.kek.jp/project/soi/seabas.html 47

Authors' Biographies

YASUO ARAI

Yasuo Arai received his Ph.D. in nuclear science from To-hoku University in 1982. Since 1982, he has been working at High Energy Accelerator Research Organization (named KEK). From 1982–1986, he worked on the data acquisition system for the VENUS experiment at the electron-positron collider accelerator TRISTAN. From 1987, he has been working on the development of readout LSI for radiation detector. Especially, he has designed TDC LSIs for the ATLAS detector, which led to the discovery of Higgs particle in 2012. In 2005, he has started a new project to develop monolithic radiation image sensor by using SOI technology, and he is a leader of the SOI pixel collaboration. He is now a professor in the electronics system group of KEK.

IKUO KURACHI

Ikuo Kurachi is a professor with High Energy Accelerator Research Organization (Japan). He holds a Ph.D. in engineering (Tokyo University of Science, Japan, 2016) and B.S. in applied physics (Tokyo University of Science, Japan, 1983). He has over 30 years of experience in semiconductor device manufacturing at OKI Electric Industry Co., Ltd. (Japan), OKI Semiconductor Co., Ltd. (Japan), and Powerchip Technology Corp. (Taiwan) where he was a leader of process integration development in DRAM, NVM, LOGIC, and sensor devices. His research interests are interface state property of $Si\text{-}SiO_2$, MOSFET reliability, and process integration technologies. He has published over 40 articles in journals and conferences.

Printed in the United States
by Baker & Taylor Publisher Services